T0135666

Localization and Posture Recognition via Magneto-Inductive and Relay-Aided Sensor Networks

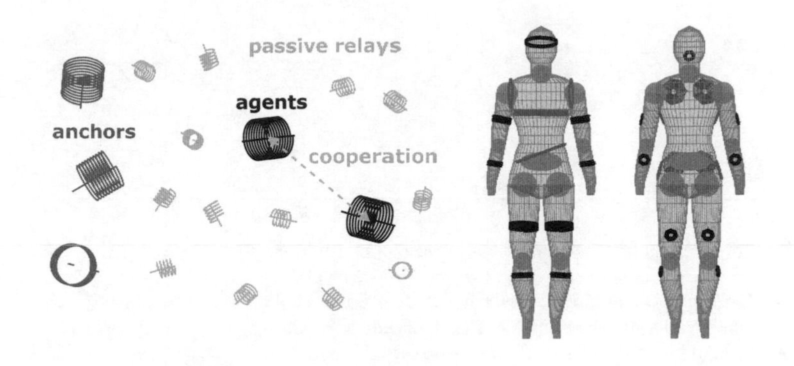

Henry Ruben Lucas Schulten

λογος

Series in Wireless Communications
edited by:
Prof. Dr. Armin Wittneben
Eidgenössische Technische Hochschule
Institut für Kommunikationstechnik
Sternwartstr. 7
CH-8092 Zürich

E-Mail: wittneben@nari.ee.ethz.ch
Url: http://www.nari.ee.ethz.ch/

Bibliographic information published by the Deutsche Nationalbibliothek

The Deutsche Nationalbibliothek lists this publication in the
Deutsche Nationalbibliografie; detailed bibliographic data are
available in the Internet at http://dnb.d-nb.de .

ISBN 978-3-8325-5483-5
ISSN 1611-2970

Logos Verlag Berlin GmbH
Georg-Knorr-Str. 4, Geb. 10, D-12681 Berlin
Tel.: +49 030 42 85 10 90
Fax: +49 030 42 85 10 92
INTERNET: http://www.logos-verlag.de

Abstract

Body-centric sensor networks play a crucial role for future eHealth systems and are envisioned to constantly monitor vitals, to provide local in-body treatments and to warn the user about dangerous conditions. The magneto-inductive physical layer, which is based on magnetic near-field coupling, has been shown to be promising for the communication of such sensor networks, primarily due to the low interaction of the human tissue with low-frequency magnetic fields. It also comprises the possibility of purely passive relaying, which is a key enabler for communication applications. However, many body-centric applications do not only require a stable connectivity and high capacity between the sensors, but also a precise knowledge of the sensors' locations and orientations. In this thesis we hence strive to extend the benefits of magneto-inductive body-centric networks to on-body and in-body localization. To this end, we first develop a theoretical framework, which is based on circuit theory and draws heavily on the Cramér-Rao lower bound. The scaling behavior of magneto-inductive localization with multiple anchors (observing infrastructure coil sensors) and agents (coil sensors that are to be localized) is analyzed and the impact of practical parameters such as the coil dimensions, the transmit power and the operating frequency is quantified.

It becomes evident that the distance dependency of the magnetic near field does not only lead to quickly decreasing channel gains, which is expected, but rather that it limits the position root-mean-square error by making it highly directional. It is this directional asymmetry, which often dominates the overall position root-mean-square error. This thesis proposes two different means to mitigate this asymmetry, improve the position root-mean-square error by orders of magnitude, and generally enhance magneto-inductive localization. The first approach uses purely passive relays (resonantly loaded coils), which provide additional signals paths from the agents to the anchors and may be exploited for the localization. The second approach introduces cooperation between all agents, which allows to share inter-agent channel state information. Both approaches are investigated by simulation and their individual drawbacks are extenuated by simple functionality extensions, such as a load-switching of the passive relays or a proper initialization for the highly-dimensional cooperative localization. Another contribution of this work is the derivation of an analytic closed-form expression for the maximum likelihood position estimator for a single pair of three-axis coils.

Based on the theoretical framework, a novel concept for a magneto-inductive human posture recognition system is proposed. This system relies on anchors that are centralized on the human torso and on purely passive coils that are placed on the extremities. Note that the approach is inherently low-power (purely passive relays) and low-complexity (low-frequency impedance measurements). The magnetic coupling between all coils, which depends on the body posture, leads to a detuning of the anchors' input impedances. The relationship between anchor input impedances and postures can consequently be learned via supervised classifiers to enable posture recognition capabilities. For this concept neither the location of the anchor coils nor of the purely passive coils need to be known in advance.

With the goal in mind to design and implement an experimental system, we study and optimize this concept via simulation for different body types, coil designs, coil placements, and also for different types of noise. For realistic noise levels, these simulations yield a classification accuracy of more than 90 %, even when only considering single-frequency impedance measurements. Based on these results an experimental system is implemented with low-cost materials. An extensive measurement campaign in a real-world office environment is conducted. It confirms the simulations and an excellent classification accuracy is achieved. However, this classification accuracy degrades substantially when unaccounted posture variations and coil displacements disturb our testing data. To increase the robustness in this practical setting, it is proposed to measure the anchor impedance at multiple frequencies. With respect to computational complexity, different types of classifiers are compared. Moreover, we consider different feature spaces for the machine learning algorithms, which have practical implications for the hardware complexity. We find that the using the magnitudes of the impedances as features (instead of the complex impedances) already leads to a satisfactory performance while enjoying the benefit of reduced complexity.

Kurzfassung

Körpernahe Sensornetzwerke spielen eine entscheidende Rolle für künftige E-Health Systeme und sollen für verschiedene Aufgabenbereiche zum Einsatz kommen. Diese Aufgaben umfassen zum Beispiel die ständige Überwachung der Vitalfunktionen, das Warnen des Nutzers vor gefährlichen Situationen oder die minimalinvasive Behandlung von Tumoren. Magnetische Induktion hat sich für die Kommunikation innerhalb solcher Sensornetzwerke hat sich als vielversprechend erwiesen, vor allem wegen der geringen Wechselwirkung des menschlichen Gewebes mit niederfrequenten Magnetfeldern. Das magnetische Nahfeld bietet ausserdem die Möglichkeit passive Spulen einzubinden, welche die Kanalkapazität erhöhen können und damit neue Anwendungen ermöglichen. Viele körpernahe Anwendungen erfordern jedoch nicht nur eine stabile Konnektivität und hohe Kanalkapazität, sondern auch eine genaue Kenntnis über die Position und Ausrichtungen aller Sensoren. In dieser Arbeit versuchen wir daher die Vorteile von magneto-induktiver Kommunikation auch auf magneto-induktive Lokalisierung auszuweiten. Zu diesem Zweck entwickeln wir zunächst ein schaltungstheoretisches Modell und betrachten die zugehörige Cramér-Rao-Ungleichung. Hierdurch sind wir in der Lage das Skalierungsverhalten der magneto-induktiven Lokalisierung mit mehreren Ankern (beobachtende Infrastruktur-Spulensensoren) und Agenten (zu lokalisierende Spulensensoren) zu analysieren und den Einfluss von praktischen Parametern wie den Spulenabmessungen, der Sendeleistung oder der Betriebsfrequenz zu quantifizieren.

Zusätzlich zeigt es sich, dass der mittlere quadratische Positionsfehler durch die Richtungsabhängigkeit des Systems dominiert wird, welche wiederum ein Resultat der starken Distanzabhängigkeit des magnetischen Nahfelds ist. Es werden folglich zwei verschiedene Methoden vorgeschlagen, welche diese Richtungsabhängigkeit abschwächen, den mittleren quadratischen Positionsfehler um Größenordnungen reduzieren, und magneto-induktive Lokalisierung andersweitig verbessern. Der erste Ansatz verwendet resonante und rein passive Spulen, die zusätzliche Signalpfade erzeugen und für die Lokalisierung ausgenutzt werden können. Der zweite Ansatz basiert auf der Kooperation zwischen allen Agenten, welche es ermöglicht Kanalzustandsinformation zwischen den Agenten für die Lokalisierung auszunutzen. Beide Ansätze werden simulationsgestützt untersucht und ihre individuellen Nachteile werden durch einfache Funktionserweiterungen gemildert. Diese Funktionserweiterungen umfassen zum Beispiel eine Umschaltung der Lastimpedanz an den passiven Spulen oder eine geeignete Initialisierung für die hochdimensionale kooperative Lokalisierung. Ein weiterer Beitrag dieser Arbeit ist die Herleitung des Maximum-Likelihood-Positionsschätzer für ein einzelnes Paar von Drei-Achsen-Spulen.

Basierend auf dem schaltungstheoretischen Modell wird ausserdem ein neuartiges Konzept für die magneto-induktive Haltungsdetektion des menschlichen Körpers vorgeschlagen. Dieses System nutzt Anker, die zentral am Torso angebracht sind, und rein passive Spulen, die an den Extremitäten platziert werden. Die magnetische Kopplung zwischen allen Spulen führt hierbei zu einer Veränderung der Eingangsimpedanzen der Anker. Die Stärke dieser Veränderung hängt von der Körperhaltung ab und die Beziehung zwischen den Impedanzen und der Körperhaltung kann folglich durch überwachte Klassifikationsalgorithmen erlernt werden. Der Ansatz ist inhärent stromsparend (rein passive Spulen) und von geringer Komplexität (Impedanzmessungen bei niedrigen Frequenzen). Für dieses Konzept muss zudem weder die Lage der Ankerspulen noch die Lage der rein passiven Spulen im Voraus bekannt sein.

Wir untersuchen und optimieren dieses Konzept durch Simulationen für verschiedene Körpertypen, Spulenstrukturen und Arten von Störungen, mit dem Endziel ein experimentelles System zu implementieren. Für realistische Störgrössen ergeben diese Simulationen eine Klassifizierungsgenauigkeit von mehr als 90 %, selbst wenn die Impedanzmessungen nur auf einer einzigen Frequenz durchgeführt werden. Basierend auf diesen Ergebnissen wird ein experimentelles System mit kostengünstigen Materialien nachgebaut. Mit diesem System wird eine umfangreiche Messkampagne in einer Büroumgebung durchgeführt. Die zugehörigen Ergebnisse zeigen eine ausgezeichnete Klassifizierungsgenauigkeit und bestätigen damit die vorherigen Simulationen. Diese Klassifizierungsgenauigkeit verschlechtert sich jedoch erheblich, wenn nicht berücksichtigte Haltungsänderungen und Spulenverschiebungen unsere Testdaten stören. Um die Robustheit gegen solche praktischen Störungen zu erhöhen, nutzen wir die Impedanz auf verschiedenen Frequenzen. Für die resultierenden hochdimensionalen Daten werden verschiedene Arten von Klassifikationsalgorithmen miteinander verglichen. Des Weiteren wird unter Anbetracht der Hardwarekomplexität untersucht, welche Messdaten den grössten Einfluss auf die Klassifizierungsgenauigkeit haben. Wir stellen fest, dass der Betrag der Impedanzen (anstelle der komplexen Impedanzen) bereits zu einer zufriedenstellenden Genauigkeit führt, trotz gleichzeitiger Reduktion der Komplexität.

Contents

Chapter 1

Motivation and Contributions

This chapter gives a brief overview of current challenges for eHealth applications and their potential for future personalized health care. It further explains why magnetic induction is a promising technology to tackle some of these challenges. Additionally, related work and its shortcomings are summarized for both localization and posture recognition. Lastly, the contributions of this dissertation are specified and all collaborations are declared.

1.1 Wireless Technologies for Body-Centric Applications

The technological advancements of the last decades allowed for a miniaturization and price reduction of powerful electronic systems. This gave rise to a widespread availability of mobile devices and sensors with currently about 6 billion mobile phone users worldwide [1] and an even larger number of **I**nternet **o**f **T**hings (IoT) connected devices [2]. The associated increase in ubiquitous computational power and sensing capabilities paves the way for smart cities, the IoT and other futuristic applications, both on an individual as well as a population level. One crucial application field on the individual level is that of eHealth, which is envisioned to enable a continuous monitoring of vitals and warn users if possibly harmful anomalies are detected [3]. In clinical settings, it is further expected that future body area networks will be a key enabler for next generation personalized health care e.g. by allowing for controllable robot swarms within the human body that may target cancerous cells or dispense medication in delimited areas [4–6]. However, some of these body-centric applications have novel requirements, which are not always be met by conventional radio technologies. In particular, they require small sensor dimensions [7], sufficient material penetration for human tissue [8], ultra-high reliability and low-energy consumption [9, 10]. Additionally, as possibly hundreds or even thousand of sensors might be required, a low unit cost is desired. The low energy requirements are of special importance for some applications, as a routine retrieval of the sensors is often unfeasible, e.g. for sensors placed within the body or clothing [11]. Moreover, the small size requirements may severely limit the volume and hence capacity of the batteries used [12]. For these cases it may be helpful to power the sensors externally via **W**ireless **P**ower **T**ransfer (WPT) or to use additional hardware that allows for continuous energy harvesting by other external means [13].

The physical layer technologies which are used differ depending on the specific task at hand. However, traditional radio, e.g. via **W**ireless **L**ocal **A**rea **N**etwork (WLAN), bluetooth, or **U**ltra-**W**ide**B**and (UWB), still remains the main contender for communication purposes on and around the human body [14, 15]. Alternative technologies such as ultrasonic transmission [16–18] or magnetic near field communication [19] have also been proposed, but so far lack a widespread adoption. For a variety of tasks, these physical layer technologies are additionally combined with local on-body or in-body sensors such as **I**nertial **M**easurement **U**nits (IMUs), pressure sensors, or various medical sensors [14]. Traditional radio technologies are advantageous as they have a low free space path loss, which is well-suited for long-range communication. They are hence well-established with an extensive existing infrastructure. Yet, their major drawbacks for body-centric application are fading, e.g. as a result of multi-path propagation, and severe shadowing caused by human tissue. Moreover, for an efficient radiation characteristic electric antennas need to be matched to the operating wavelength. This requirement limits their efficiency as micro sensors unless ultra-high operating frequencies are used, which in turn reduces the material penetration and complicates the synchronization between sensors. These drawbacks still allow for communication on and around the body, but make it challenging to use radio as main mechanism for in-body communication or sensing applications [20,21]. Ultrasonic wireless body area networks on the other hand incur a reduced path loss compared to radio systems and are thus already established in challenging environments, e.g. for long-range underwater networks [22]. However, they are still affected by undesired reflections and scattering. Moreover, the operating frequency and transmission power have to be chosen carefully to obtain a sufficient channel capacity without triggering health issues through adverse heating of tissue or cavitations [17].

Magnetic induction and more specifically AC-based low-frequency magnetic induction has the following advantages for communication and localization purposes: It has a high material penetration and is hence robust to the limiting attenuation of otherwise challenging materials such as human tissue or water [23, 24]. It has a low interaction with the environment and can be predicted well by simple analytic models [25,26]. The associated antennas operate in the near field and have a low and almost purely ohmic resistance. In combination with a resonant multi-turn coil design, this low resistance allows for a high current that can achieve significant channel gains and wireless power transfer over a short range of interest [27]. As radiation is undesired, the antennas do not require a length of about $\frac{\lambda}{2}$ and the overall sensors can come in various sizes, i.e. they can easily be scaled from meters to micrometers [28]. Further, the received power P decays according to $P \propto d^{-6}$ due to magnetic near field path loss, with distance d between transmitter and receiver. This makes **M**agneto-**I**nductive (MI) networks highly susceptible to variations of the sensor positions, which is beneficial for the obtainable localization accuracy. It also leads to a so-called *magnetic bubble* around the antenna [29], i.e. the strong decay limits interference between neighboring systems which enables spatial reuse and enhances security [28]. The low frequency operation additionally simplifies the phase synchronization, and even the time synchronization requirements are relaxed e.g. compared to ultra-wideband technology. These simplifications combined with the low production cost of the involved sensors and other associated hardware make many MI systems low-cost and low-complexity. Moreover, magnetic induction allows for the use of purely passive relays which may further enhance the channel gain and have shown to be useful for both communication [30–32] and wireless power transfer [33]. Compared to passive tags which are known from backscatter communication [34], these purely passive relays do not require an **I**ntegrated **C**ircuit (IC) or have to rely on load modulation in order to be beneficial. At a first glance, magnetic induction hence satisfies all mentioned requirements for body-centric applications. However, the sextic distance-dependency of the path loss over also limits their applicability, since small sensors cannot reliably communicate if their separation is large relative to their antenna sizes. In addition, the low operating frequency reduces the usable bandwidth and hence the overall achievable rate, which complicates the use of data-intensive applications such as high-resolution video transmissions. Moreover, MI systems are not well-established, so despite their low individual cost they may require larger initial investments. Lastly, the magnetic near field can be drastically impacted by nearby conductors [35]. Nevertheless, MI systems pose as promising candidate for future body-centric networks and applications, especially for in-body systems. It is thus of interest to further advance this type of technology and possibly mitigate some of its drawbacks. In this work in particular, we study and contribute to the current state of the art in two related fields: *MI localization* and *MI posture recognition.*

1.2 Localization in Sensor Networks

Many of the envisioned applications do not only require communication capabilities from a sensor network, but also knowledge on the spatial positions of the sensors. This knowledge can be used to enhance the existing communication links or to use the network more efficiently [36,37]. On top of this, some tasks such as access control or robot navigation may not function at all without either the full spatial information or at least notions of proximity within the network [38]. While **G**lobal **N**avigation **S**atellite **S**ystems (GNSSs) are established as the gold standard for localization in outdoor environments, they lack the accuracy that body-centric applications require. Furthermore, the accuracy of GNSS localization degrades even further when used in indoor scenarios or other challenging environments. Alternative localization approaches are e.g. based on radio, inertial measurement units, or camera systems, but issues such as multipath fading, drift errors or the need for mobility hinder these technologies, and so far no *silver bullet* has been found [39]. Localization using magnetic near fields has also been studied extensively [39–41], e.g. for indoor localization or in-body localization. Generally, it can be distinguished between systems which use ambient magnetic fields such as the geomagnetic field [42,43], static magnetic fields e.g. via permanent magnets [44–47], and those that use dedicated AC-generated magnetic fields [31,35,48–54]. The geomagnetic approaches use the unique and robust distortion patterns of the earth's magnetic field, which occur as a result of nearby construction materials. Via initial calibration, a fingerprinting database with these distortions can be created which associates each field intensity with a certain position. During operation, new field intensity measurements are then re-associated with the saved positions to localize a user. Since many devices such as smartphones phones equipped with three-axis magnetometers by default, this approach does not require additional hardware making it easy to use. However, recent surveys [40,41] found that the effort of generating such a database is significant and the resulting accuracy is often lacking due to the low differences in magnetic field intensity, which range from 25 µT to 65 µT on the earth's surface [55]. In [43] it was proposed to mitigate the latter issue by relying on multiple consecutive measurements for the fingerprinting database and to additionally incorporate opportunistic WLAN signals to boost accuracy and allow for tracking. This combination resulted in an overall localization error of less than 5 m for 80 % of cases when tested in large parking lots or supermarkets. Yet, both the required generation of a fingerprinting database as well as the lack of accuracy make localization via the ambient geomagnetic field unsuitable for body-centric applications. Moreover, even with dedicated magnetic fields the generation of a fingerprinting database inside the human body would be a challenging task and may require imaging or other additional means of localization. For similar tasks, such as capsule endoscopy or minimally-invasive surgery, it is beneficial to have a strong and dedicated field, which can also be modeled analytically [56]. In [44,47] the authors hence use small permanent magnets as sources and an array of surrounding hall sensors for measurements. They estimated the transmitters position and orientation by numerically minimizing the difference between the measured field strength and its expected noiseless value according to the analytic model of the path loss, e.g. by using the Levenberg-Marquardt algorithm. In these works, the obtained average localization errors were in the millimeter regime, which varied depending on the network dimensions, the sensor topology, the size of the involved permanent magnets, the employed minimization algorithms, the type of measurement sensors and other parameters. While permanent magnets have the advantage of not requiring a dedicated transmit current, the resulting coverage area and the obtainable peak accuracy of these systems are lower than those of AC-driven MI systems of equal dimensions [40]. AC-driven MI localization systems also commonly deploy an analytic path loss model to estimate the position and orientation of the involved coil antennas. Due to their flexibly adjustable coverage area that only requires a scaling of the coil antennas and the transmit power, they have been studied for outdoor localization in harsh environments [51–53,57,58], for indoor localization [25,26,59–61], and for in-body localization [28,62,63]. Moreover, for three-axis coils, empirically motivated closed-form position estimators have been derived in [25] allowing for a lower computational complexity compared to the numerical minimization approaches. Additionally, the localization of passive resonant coil antennas that do not require a power source has been studied in [52,64]. In [65] it has also been demonstrated that the load-switching of such passive coils can have further benefits for MI localization. In particular, it was shown that this load-switching leads to multiple independent measurements, which can resolve position ambiguities at a single measurement coil, therewith enabling single-anchor localization. While the mentioned MI systems show promise and even found their way to commercial use [66,67], they still suffer drastically by the range limitations of the magnetic near field. That is, if the required coverage area is large compared to the coil dimensions, even the AC-driven MI systems become quickly unreliable. Additionally, metal-rich environments cause significant distortions and make those systems unsuitable if being unaccounted for [25].

1.3 Posture and Activity Recognition

With accurate location knowledge of multiple fixed sensors on a human body, it is possible to estimate the corresponding body posture or movement. This is done by matching a simplified joint-based human body model to the previously estimated sensor locations [68, 69]. Alternatively, the joint-based body model may also be matched to full body parts which are detectable via camera images [70] to bypass the need for on-body sensors. Such human posture or activity estimation is used for movie animations [69] or for navigation in virtual reality [71]. However, instead of fully estimating the human posture, it can be sufficient to only classify a user's current posture from a finite set of possible options by relying on supervised classifiers, which are trained on previously recorded data. Posture recognition or classification allows for drastically reduced system complexity, as the individual sensor or body part positions do not need to be known. This simpler approach may be crucial for individual health care in terms of injury prevention or rehabilitation [72–74], but other use cases such as fitness tracking or sports coaching may also benefit from it [75]. More particular, reliable posture classification may prevent **M**usculo**S**keletal **D**isorders (MSDs), which are the second leading cause for disability and are often caused by over-extensions or a repetition of unhealthy postures and movement patterns. These disorders are not only detrimental to the economy on a global scale, they also cause chronic pain and reduce the overall life expectancy [76]. Posture recognition systems which can be worn comfortably on a daily basis may hence be used to recognize such unhealthy behavior, intervene by warnings or stimuli, and consequently prevent the emergence of these disorders in the first place [74]. The current gold standard when it comes to posture recognition and classification is offered by vision-based systems, which can nowadays operate without optical markers [70, 72, 77]. Yet, these vision-based systems are immobile, require a line-of-sight and are often associated with a high cost. Current wearable solutions that do not require external fixed infrastructure are e.g. based on radio [73,78], inertial measurement units [79–85], strain sensors [11,86,87], or magnetic near-field systems [29,88]. These wearable systems however share the common issue of needing distributed power sources on or wired connections to all involved nodes. As a result, current systems are often designed to only cover and monitor a single body part of interest such as the spine [85,87]. For the recognition of full body postures, the need of active nodes reduces practicability and widespread adoption, as it requires the nodes to either be charged periodically or to use specific full body clothing.

1.4 Contributions, Publications, and Structure

Sec. 1.1 to Sec. 1.3 summarized why low-frequency magnetic induction is a promising physical layer technology for body-centric tasks. Nevertheless, some underlying issues such as a significant range limitation hinder its widespread adoption, e.g. for envisioned eHealth applications. This thesis hence strives to broaden our understanding of MI networks by studying challenges and opportunities of both MI localization and MI posture detection. In detail, we make the following core contributions to extend the current state of the art:

- We provide simple approximations which characterize how the scaling of key network parameters affects the MI localization accuracy. We further identify common issues that degrade the localization accuracy, such as an asymmetry of the associated **C**ramér-**R**ao **L**ower **B**ound (CRLB).
- We investigate the impact of passive relays on MI networks and reveal their versatile use for localization. We also explain possible drawbacks that occur when passive relays are employed for random network topologies and provide ways to counteract them.
- We study cooperation within MI networks and discuss the resulting trade-off between localization accuracy and computational complexity.
- We derive closed-form solutions of the **M**aximum **L**ikelihood (ML) position, orientation, and distance estimators for a single pair of three-axis coils.
- We propose a novel concept to classify the topology of a MI network without knowledge of the coils' deployments. This concept only requires impedance measurements on a few active coils while all other coils in the network are purely passive.
- We demonstrate the feasibility of low-complexity and low-cost MI posture recognition by simulation and experiment. We identify key parameters for posture recognition, compare the applicability of different supervised classifiers, and study means to improve robustness against domain shifts.

Some of these contributions were already published by us at international conferences or as patent application. This dissertation is hence generally based on or related to the following publications:

Papers

- [89] **H. Schulten** and A. Wittneben, "Robust Multi-Frequency Posture Detection Based on Purely Passive Magneto-Inductive Tags," in ICC 2022 - IEEE International Conference on Communications. IEEE, May 2022, pp. 1–6.
- [90] **H. Schulten** and A. Wittneben, "Experimental Study of Posture Detection Using Purely Passive Magneto-Inductive Tags," in WCNC 2022 - IEEE Wireless Communications and Networking Conference. IEEE, Apr. 2022, pp. 1–6.
- [91] **H. Schulten**, F. Wernli, and A. Wittneben, "Learning-Based Posture Detection Using Purely Passive Magneto-Inductive Tags," in Globecom 2021 - IEEE Global Communications Conference. IEEE, Dec. 2021, pp. 1–6.
- [92] **H. Schulten**, G. Dumphart, A. Koskinas, and A. Wittneben, "Cooperative Magneto-Inductive Localization," in PIMRC 2021 - IEEE 32nd Annual International Symposium on Personal, Indoor and Mobile Radio Communications. IEEE, Sep. 2021, pp. 1–7.
- [93] **H. Schulten** and A. Wittneben, "Magneto-Inductive Localization: Fundamentals of Passive Relaying and Load Switching," in ICC 2020 - IEEE International Conference on Communications. IEEE, Jun. 2020, pp. 1–6.
- [61] G. Dumphart, **H. Schulten**, B. Bhatia, C. Sulser, and A. Wittneben, "Practical Accuracy Limits of Radiation-Aware Magneto-Inductive 3D Localization," in ICC 2019 - IEEE International Conference on Communications Workshops. IEEE, May 2019, pp. 1–6.
- [38] R. Heyn, M. Kuhn, **H. Schulten**, G. Dumphart, J. Zwyssig, F. Trosch, and A. Wittneben, "User Tracking for Access Control with Bluetooth Low Energy," in VTC2019- Spring - IEEE 89th Vehicular Technology Conference. IEEE, Apr. 2019, pp. 1–7.
- [94] **H. Schulten**, M. Kuhn, R. Heyn, G. Dumphart, F. Trosch, and A. Wittneben, "On the Crucial Impact of Antennas and Diversity on BLE RSSI-Based Indoor Localization," in VTC2019-Spring - IEEE 89th Vehicular Technology Conference. IEEE, Apr. 2019, pp. 1–6.

Patent Applications

- [95] **H. Schulten**, and A. Wittneben, "Method and Apparatus for Determining a Spatial Configuration of a Wireless Inductive Network and for Pose Detection," European Patent Request EP21 211 962.2, Dec., 2021.
- [96] F. Trosch, A. Wittneben, **H. Schulten**,, J. Zwyssig, and M. Kuhn, "Zugangskontrollsystem und Verfahren zum Betreiben eines Zugangskontrollsystems," World Patent WO2 020 216 877A1, Oct., 2020.

This thesis is structured as follows:

After summarizing the theoretic background in Cpt. 2, we introduce a circuit-based signal model for MI coupling in Cpt. 3, which works with a multitude of arbitrarily arranged active and passive coil antennas. We derive the CRLB on the position **R**oot-**M**ean-**S**quare **E**rror (RMSE) for this model (cf. Appendix A) and introduce a measure of its spatial asymmetry.

In Cpt. 4 we draw on this bound to approximate the scaling characteristics of MI localization with respect to common design parameters. We further examine differences between the ranging of active and passive sensors. Additionally, it is studied how the localization capabilities are degraded due to a mutual detuning of the coil antennas in dense networks. Lastly, localization bottlenecks inherent to the use of the magnetic near field, and more specifically its severe distance dependency, are identified for specific network constellations.

In Cpt. 5, which incorporates our work from [93], we introduce passive tags to the network. These tags are additional coil antennas equipped with a fixed or variable load. The impact of these nodes on the localization is examined and their many advantages for MI localization are summarized. However, adverse effects occur in case of close proximity between those tags and the sensor nodes that are to be localized. It is hence further investigated how switching the passive tag loads may be utilized, not only to mitigate the adverse effects but also to further boost the localization accuracy and range.

In Cpt. 6, which is an extension of [92], we conduct a study of cooperative localization for magneto-inductive sensor networks. Cooperation is defined as the agents' ability to measure all inter-agent channel gains and forward this information to the anchors. The impact of the number of cooperating nodes is studied by means of numerical optimizers, both for solenoid and three-axis coil antennas. For a single pair of three-axis coils, we further derive a closed-form solution for the ML position and orientation estimator based on channel gain information between all subcoils (cf. Appendix B). Lastly, we characterize the trade-off between the achieved cooperation gain for localization and the associated increased computational complexity for the numerical optimizer.

In Cpt. 7 we shift attention away from estimation-based MI localization and propose a novel concept to recognize the entire topology of a MI network. This concept is motivated by magnetic near field fingerprinting and uses magnetic field distortion caused by resonant passive coil antennas. These intentionally caused distortions change for different network topologies and can be measured by means of single-frequency complex impedances. Learning the relationship between a topology and the resulting complex impedances via training data allows for re-association and hence the recognition of the network topology. The proposed approach inherently low-power and low-complexity.

In Cpt. 8 we expand our work from [91] and use the topology recognition concept proposed in Cpt. 7 to realize a MI posture recognition system. In detail, we place measuring coil antennas on the human torso and passive coil antennas on the extremities, such that everyday posture theoretically leads to different observed impedance values. Yet, it is unclear whether these theoretic changes are significant enough to make the postures distinguishable when measurement noise is present and when respecting the confinements of the human body. We hence study the approaches feasibility via simulations and analyze the impact of different types of noise as well as practical perturbations. We further compare the obtained classification accuracy for different coil antennas, body models, and supervised classifiers.

One design of the posture recognition system which was investigated in Cpt. 8 is then implemented with low-cost materials. In Cpt. 9, which is based on [89] and [90], we experimentally test this demonstrator system in an office environment. We study it for a worsening noise characteristic, which would be expected for wearable low-cost impedance analyzers. We also compare supervised classifiers to the theoretical ML classifier and examine how minor variations of the postures and coil displacements affect the classification. The resulting issues are addressed by extending the system to use multi-frequency impedance measurements.

Lastly, Cpt. 10 provides a short overview on the most important results, both on a qualitative as well as a quantitative level.

1.5 Acknowledgments and Joint Work

Whenever I hear about the phrase *the self-made man or woman*, I cannot help myself but wonder, who out there in this world has truly achieved great things without the support and influence of others. As I am no exception to this consideration, I am thankful to a great many people without whom the creation of this work would not have been possible.

First, I want to declare the following technical contributions and collaborations: The works of Eric Slottke [31] and Gregor Dumphart [28], which are based on publications by Michel Ivrlak and Josef Nossek [97], laid the foundation for the closely related signal and noise model of Cpt. 3. Vincent Wüst assisted me with a simulative investigation of passive links that paved the way for Sec. 4.1. The results of Cpt. 6 benefited from simulations by Antonios Koskinas and discussions with Gregor Dumphart. Florian Wernli and Viviane Marty contributed to Cpt. 8 and Cpt. 9 by running advised simulations. Lastly, Florian Wernli, Bharat Bhatia, and Michael Lerjen assisted with the hardware design and measurement campaigns, which were required for Cpt. 9.

On a personal note, I am grateful to my supervisor Armin Wittneben, who not only gave me the chance of pursuing my graduate studies in the first place, but also taught me about wireless communications, the importance of scientific rigor, and who intuitively steered me in the right directions whenever I needed guidance. Second, I would like to thank Marc Kuhn for many important hints and his co-supervision of this work. Moreover, I am thankful to the Christoph Mecklenbräuker for acting as a referee to this dissertation. I would also like to express my gratitude to Gregor Dumpart for many invaluable discussions and our collaborations, Robert Heyn for providing an alternative perspective and good companionship, and other former colleagues for their assistance and for creating an inspiring work environment. Lastly, I am also forever grateful to my family and friends. To my mother Julia for encouraging me to pursue what I am passionate about and for teaching me to see opportunity rather than failure. To my other family members and loyal friends for distracting me when it was needed and for making sure I do not take myself too seriously all the time. To Frederike, who made me aware of my true goals in life, who taught me to find joy in unexpected places, and whose company never fails to put a smile on my face.

Chapter 2

Coil Antennas and Magneto-Inductive Coupling

The entire thesis is based on sensors which use MI coupling as primary propagation mechanism. To establish such wireless links, the sensors deploy so called coil antennas. Various works such as [27–29,31,52] already summarize common models for coil antennas and offer circuit-based representations to analytically model the MI interactions of such coils. Yet, for convenience this chapter briefly summarizes the most important aspects of these models and explicates the underlying restrictions in a systematic manner. That is, Sec. 2.1 states the main assumptions under which this thesis operates and introduces our modeling approaches for coil antennas. In Sec. 2.2 the general multi-port coupling approach is presented, which describes the magnetic near-field interactions between multiple such coils.

2.1 Inductive Coupling and Limiting Assumptions

Generally, the electromagnetic interaction between all sensors can be described adequately by the four famous and well-established equations of Maxwell [98,99]. In plain words, they describe that (i) charges produce an electric field that diverges from positive to negative charges, (ii) the magnetic field has no such sources or sinks, (iii) a time-varying magnetic field generates solenoidal electric fields, and (iv) either electric currents or time-varying electric fields create solenoidal magnetic fields. Notion (iii) is also known as Faraday's law of induction and together with (iv), also called Ampere's law, establishes the basis for the herein considered wireless links. Yet, evaluation of the Maxwell equations for practical problems is often unfeasible. It is hence common practice to either resort to numerical approximations with finite elements [100] or to use simplifying assumptions, which in turn drastically reduce the complexity of the equations. In this work we do the latter, namely, we assume that for the inductive interaction between different sensors it holds that:

1. All considered signals are harmonic with a radial operating frequency $\omega = 2\pi f$, where f is the ordinary frequency.

2. The **M**agneto**Q**uasi**S**tatic (MQS) approximation applies, i.e. the generation of solenoidal magnetic fields due to displacement currents can be neglected. We can hence ignore retardation and residual radiation [101,102]. This is approximately the case if all signals have a low-frequency and hence cause low-frequency fields with a corresponding wavelength λ that is significantly larger than the distance between any pair of considered conductors [28].

3. All considered conductors are infinitely thin-wired, so the current distribution along the cross section of all wires is one dimensional.

4. All conductors used are electrically small, i.e. $l^{\text{wire}} \leq \frac{\lambda}{10} \ll \lambda$ with l^{wire} as full wire length. As a result, we can assume the one-dimensional current distribution over the entire length of the conductor to be constant [27,103].

5. All propagation media are linear, isotropic, and homogeneous.

Next we consider a transmit conductor m with current i_m and surface enclosing boundary $\mathcal{C}_m$, as well as a receive conductor with surface enclosing boundary $\mathcal{C}_n$. Let the receive conductor have two terminals, i.e. it is non-closed and has small gap that is part of $\mathcal{C}_n$. Applying Faraday's law of induction in conjunction with Stokes' theorem, Ampere's law and the above assumptions, it is found that the induced voltage v_n across the terminals of the receive conductor (from positive to negative pole) is given by [28,31]

$$v_n = j\omega M_{m,n} i_m \tag{2.1}$$

$$M_{m,n} = M_{m,n}^{\text{Neu}} = K^{\text{Neu}} \oint_{\mathcal{C}_m} \oint_{\mathcal{C}_n} \frac{\mathbf{dl}_m \, \mathbf{dl}_n}{d^{\text{wire}}(l_m, l_n)} \,. \tag{2.2}$$

$M_{m,n}$ is called the mutual inductance, $j\omega M_{m,n}$ is referred to as mutual impedance and (2.2) is commonly known as Neumann formula [104]. The integrals are evaluated over the two conductor boundaries via their one-dimensional lengths' l_m, l_n. These length are used to define the positions $\mathbf{p}_m^{\text{wire}}(l_m)$, $\mathbf{p}_m^{\text{wire}}(l_n)$ of each infinitesimal wire element on these structures and $d_{m,n}^{\text{wire}}(l_m, l_n) = \|\mathbf{p}_m^{\text{wire}}(l_m) - \mathbf{p}_n^{\text{wire}}(l_n)\|$ describes all pairwise distances between these elements. Moreover, $\mathbf{dl}_m$ and $\mathbf{dl}_n$ are the infinitesimal vectorial line-segments at each corresponding wire element position. Lastly, the multiplicative constant is given by $K^{\text{Neu}} = \frac{\mu_0 \mu_{\text{r}}}{4\pi} = \frac{\mu}{4\pi}$ with μ_0 and μ_{r} as vacuum permeability and relative permeability of the propagation medium, respectively. The analytic closed-form evaluation of the Neumann formula is only feasible for specific structures and alignments of the conductors,. However the formula can easily be calculated by numerical means. Alternatively, the conductors may sometimes also be approximated by magnetic dipoles which allows for a simpler and analytic closed-form description of the mutual inductance (cf. (2.3)).

2.2 Electrically Small and Thin-Wired Coils

A common shape of conductors used for MI networks is that of a coil antenna and more specifically that of a **S**ingle-**L**ayer **S**olenoid (SLS) coil antenna. To fully describe the location of such coils, we use the three-dimensional Cartesian coordinates $\mathbf{p} = [p_x, p_y, p_z]^\mathrm{T}$ of the coil's center and three Euler angles $\boldsymbol{\phi} = [\alpha, \beta, \gamma]^\mathrm{T}$ that describe the coil's orientation via z-y-z rotation [105] (cf. (A.19)). We summarize those variables as six-dimensional deployment vector $\boldsymbol{\psi} = [\mathbf{p}^\mathrm{T}, \boldsymbol{\phi}^\mathrm{T}]^\mathrm{T}$.

If we consider a pair m, n of such coils with sufficiently large separation $d_{m,n} = \|\mathbf{p}_m - \mathbf{p}_n\|$ relative to their own dimensions, the fields they generate may be approximated by that of an magnetic dipole which leads to their mutual inductance being expressible by [28, Eq. 2.26]

$$M_{m,n} \approx M_{m,n}^{\mathrm{Dip}} = \frac{K_{m,n}^{\mathrm{Dip}}}{d_{m,n}^3} \underbrace{\mathbf{o}_m^\mathrm{T} \overbrace{\left(\frac{3}{2}\mathbf{u}_{m,n}\mathbf{u}_{m,n}^\mathrm{T} - \frac{1}{2}\mathbf{I}_3\right)}^{\mathbf{F}_{m,n}} \mathbf{o}_n}_{J_{m,n}}, \tag{2.3}$$

where the direction vector $\mathbf{u}_{m,n} = \frac{\mathbf{p}_m - \mathbf{p}_n}{d_{m,n}}$ between both coil centers is also called direction of departure. Moreover, $\mathbf{o} = [\cos(\alpha)\sin(\beta), \sin(\alpha)\sin(\beta), \cos(\beta)]^\mathrm{T}$ represents the Cartesian unit vector of a coil's cylindrical main axis. It offers a simplified description of the coil's orientation since the parameter γ has no impact due to a dipole being rotation symmetric around its dipole axis. These orientations further determine the alignment factor of the coil pair $J_{m,n}$ with $|J_{m,n}| \leq 1$ [106, 107]. Lastly, the constant $K_{m,n}^{\mathrm{Dip}} = \frac{\mu_0 \mu_r \pi}{32} N_m^{\mathrm{coil}} N_n^{\mathrm{coil}} (D_m^{\mathrm{coil}} D_n^{\mathrm{coil}})^2$ depends on the coils' diameters D_m^{coil}, D_n^{coil} and the numbers of coil windings N_m^{coil}, N_n^{coil}. However, for closer coil separations $d_{m,n}$ this model is highly inaccurate and the numerical evaluation via Neumann formula may be required. For illustrative purposes, we show a single layer solenoid coil in its initial position (i.e. $\mathbf{p} = \mathbf{0}_{3\times1}$ and $\boldsymbol{\phi} = \mathbf{0}_{3\times1}$, which leads to $\mathbf{o} = [0, 0, 1]^\mathrm{T}$) in Fig. 2.1.

For a randomly placed or dynamically moving pair of solenoid coils, the coils can be misaligned, which leads to small values of $J_{m,n}$. Such a misalignment may hinder or even prevent magnetic induction. Three-axis coils, which are a combination of three

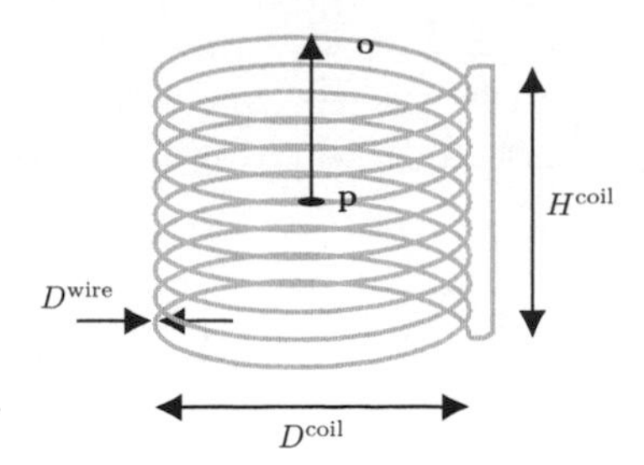

Figure 2.1: Illustration of a SLS coil antenna with $N^{\text{coil}} = 10$ windings in its initial position. Other defining parameters are also displayed.

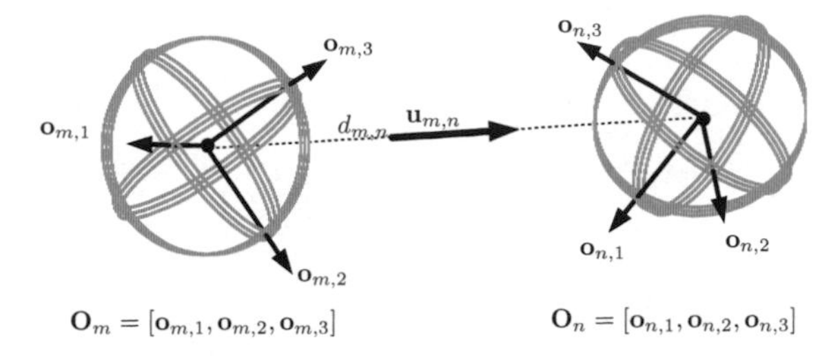

Figure 2.2: A pair (m, n) of three-axis coils, each comprising three orthogonal solenoid subcoils. The figure is adopted from [28, Fig. 3.8].

orthogonal solenoid subcoils, are hence a common design choice to mitigate such misalignment losses. An exemplary three-axis coil pair with its corresponding parameters is illustrated in Fig. 2.2. The mutual inductance matrix $\mathbf{M}^{\text{Dip}}_{m,n} \in \mathbb{R}^{3\times3}$ between such a pair of three-axis coils is the straightforward extension of (2.3) and can be expressed by [28]

$$\mathbf{M}^{\text{Dip}}_{m,n} = \frac{K^{\text{Dip}}_{m,n}}{d^3_{m,n}} \, \mathbf{O}^{\text{T}}_m \left(\frac{3}{2} \mathbf{u}_{m,n} \mathbf{u}^{\text{T}}_{m,n} - \frac{1}{2} \mathbf{I}_3 \right) \mathbf{O}_n \, , \tag{2.4}$$

with orientation matrix $\mathbf{O}_m = [\mathbf{o}_{m,1}, \mathbf{o}_{m,2}, \mathbf{o}_{m,3}]$ and analogous definition for $\mathbf{O}_n$. This extension is apparent if we e.g. look at a single element of this matrix $[\mathbf{M}^{\text{Dip}}_{m,n}]_{k,l} = \frac{K^{\text{Dip}}_{m,n}}{d^3_{m,n}} \; [\mathbf{O}^{\text{T}}_m]_{k,:} \left(\frac{3}{2} \mathbf{u}_{m,n} \mathbf{u}^{\text{T}}_{m,n} - \frac{1}{2} \mathbf{I}_3 \right) [\mathbf{O}_n]_{:,l} = M^{\text{Dip}}_{k,l}$, which corresponds to the mutual inductance between the solenoid subcoils k and l.

In circuit theory, SLS coil antennas are typically modeled by a combination of resistances, inductances, and capacitances. One of the simplest but commonly used representations is shown in Fig. 2.3 and only requires one of each such element. This

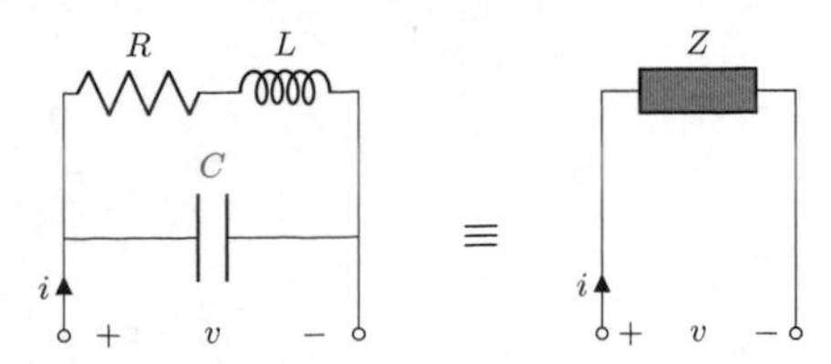

Figure 2.3: Equivalent RLC circuit representation of a coil and corresponding summarized impedance element.

model is valid as long as the operating frequencies are significantly lower than the coil's true self-resonance frequency [108]. For the remainder of this work except for Fig. 2.4, this RLC representation is visually simplified by the depiction of its corresponding equivalent complex impedance element (gray box), as indicated.

The shown inductance L of this model represents the coils' self-inductance, i.e. its tendency to oppose the change of its current, which is equal to the ratio between current-generated magnetic flux and the current itself,

$$L = \frac{\Phi(i)}{i} . \tag{2.5}$$

It may also be approximated by the double integral of (2.2) over the same coil if overlapping infinitesimal elements are excluded and an additional correcting term is added [109], so

$$L = \left(K^{\mathrm{Neu}} \oint_C \oint_{C'} \frac{\mathrm{d}\mathbf{l}\,\mathrm{d}\mathbf{l}'}{d^{\mathrm{wire}}(l, l')} \right)_{2d^{\mathrm{wire}}(l,l') > D^{\mathrm{wire}}} + K^{\mathrm{Neu}} l^{\mathrm{coil}} Y , \tag{2.6}$$

with $Y = \frac{1}{2}$ for a homogeneous current distribution across the wire diameter and $Y = 0$ if the current is only on the wire surface. Further correcting terms may be required depending on the exact coil shape but are often negligible for $2l^{\mathrm{coil}} \gg D^{\mathrm{wire}}$. This formula was originally proposed for wired loops but also reasonably extends to other coil structures.

The capacitance used illustrates an unwanted self-capacitance of the coil that is theorized to be the result of adjacent coil windings or other effects that can be explained by power line theories [110]. A common choice to evaluate the self-capacitance of a coil was given in [111]. However, this choice was recently reanalyzed, criticized and ultimately extended to the following empirically-corrected polynomial approximation for single layer solenoids [110]

$$C = \frac{4\epsilon_0\epsilon_{\mathrm{r,out}}}{\pi} H^{\mathrm{coil}} \left(1 + k_c \frac{1 + \frac{\epsilon_{\mathrm{r,in}}}{\epsilon_{\mathrm{r,out}}}}{2}\right) \left(1 + \left(\frac{H^{\mathrm{coil}}}{\pi D^{\mathrm{coil}} N^{\mathrm{coil}}}\right)^2\right), \tag{2.7}$$

$$k_c = 0.717439 \left(\frac{D^{\mathrm{coil}}}{H^{\mathrm{coil}}}\right) + 0.933048 \left(\frac{D^{\mathrm{coil}}}{H^{\mathrm{coil}}}\right)^{\frac{3}{2}} + 0.106 \left(\frac{D^{\mathrm{coil}}}{H^{\mathrm{coil}}}\right)^2 \tag{2.8}$$

where ϵ_0 is the vacuum permittivity, $\epsilon_{\mathrm{r,in}}$ is the relative permittivity of the coil core and $\epsilon_{\mathrm{r,out}}$ is the relative permittivity outside of the coil. Moreover, H^{coil} is the coil height.

Lastly, the resistance R represents a combination of various losses that occur in a coil antenna. It usually comprises ohmic losses R^{ohm} as well radiative losses R^{rad}. However, as we operate in a regime where the MQS assumption applies, the radiative losses are significantly smaller than the ohmic losses and will therefore be neglected. The dominant ohmic losses consider the wire material's resistivity, the skin effect, and the proximity effect of the coil. That is,

$$R = R^{\mathrm{ohm}} + R^{\mathrm{rad}} \approx R^{\mathrm{ohm}} = \frac{l^{\mathrm{wire}}}{\sigma^{\mathrm{wire}} \pi \delta (D^{\mathrm{wire}} - \delta)} \left(1 + \frac{R_{\mathrm{p}}}{R_0}\right) \tag{2.9}$$

with σ^{wire} as conductivity of the wire and $l^{\mathrm{wire}} = l^{\mathrm{spiral}} + l^{\mathrm{conn.}}$ as total wire length of the coil, comprising both the length of the coil's spiral part l^{spiral} and the length of the coil's connector $l^{\mathrm{conn.}}$. For solenoid coils, we choose them to be $l^{\mathrm{spiral}} = \sqrt{(\pi D^{\mathrm{coil}} N^{\mathrm{coil}})^2 + (H^{\mathrm{coil}})^2}$ and $l^{\mathrm{conn.}} = 2\frac{H^{\mathrm{coil}} - D^{\mathrm{wire}}}{N^{\mathrm{coil}}} + H^{\mathrm{coil}}$. Moreover, $\delta = \min\left\{\frac{D^{\mathrm{wire}}}{2}, \sqrt{\frac{2}{\omega\mu\sigma^{\mathrm{wire}}}}\right\}$ is the skin depth and $\frac{R_{\mathrm{p}}}{R_0}$ is the proximity-based increase of the ohmic losses. More details and values for the latter can be found in [27] for different coil parameters.

With this representation, the overall frequency-dependent coil impedance evaluates to

$$Z = (R + j\omega L) \parallel \frac{1}{j\omega C} = \frac{R + j\omega L}{1 + j\omega C\,(R + j\omega L)}. \tag{2.10}$$

Moreover, the coils self-resonance frequency f^{self} follows as [28, Eq. 2.37]

$$f^{\mathrm{self}} = \frac{1}{2\pi}\sqrt{\frac{1}{LC} - \left(\frac{R}{L}\right)^2} \approx \frac{1}{2\pi}\frac{1}{\sqrt{LC}}, \tag{2.11}$$

where the direct evaluation is complicated by the frequency-dependency of the resistance and the latter approximation is hence commonly used for coils with a high quality factor Q that is defined as [28, Eq. 2.38]

$$Q = \frac{2\pi f^{\text{res}} L}{R} = \frac{1}{2\pi f^{\text{res}} RC}, \tag{2.12}$$

where f^{res} may be the coil's native self-resonance frequency or a resonance frequency that was obtained by connecting additional impedances or circuitry.

2.3 Inductive Coupling in a Network of Coil Antennas

Having established a model for individual coil antennas, we next look at the magnetic near-field coupling between N coil antennas. Clearly, the voltage at each coil antenna's port is the result of its own self-induced voltage superposed by the induced voltages due to all other coil antennas. This relationship can be modeled conveniently by the multiport representation illustrated in Fig. 2.4, which was also deployed by [28, 31, 97, 112]. The current-voltage relationship between all coils is hence be characterized by the complex impedance matrix $\mathbf{Z}_\text{C}$ of the entire antenna multiport. This impedance matrix depends on the respective self-inductances L_n, the summarized ohmic and radiative losses R_n, and the self capacitances C_n of each coil $n = 1, \ldots, , N$, as well as the mutual inductances $M_{m,n} = M_{n,m}$ between any two different coils m and n with $m = 1, \ldots, N$ and $m \neq n$. In detail, we find [19]

$$\mathbf{v} = \mathbf{Z}_\text{C}\, \mathbf{i}\,, \tag{2.13}$$

$$\mathbf{Z}_\text{C} = \mathbf{Z}_\text{C}^0 \parallel \frac{1}{j\omega\mathbf{C}} = \left((\mathbf{Z}_\text{C}^0)^{-1} + j\omega\mathbf{C}\right)^{-1}, \tag{2.14}$$

where $\mathbf{v} = [v_1, \ldots, v_N]^\text{T}$, $\mathbf{i} = [i_1, \ldots, i_N]^\text{T}$, and $\mathbf{C} = \text{diag}(C_1, \ldots, C_N)$. The matrix $\mathbf{Z}_\text{C}^0 = (\mathbf{R} + j\omega\mathbf{M})$ is the antenna impedance matrix without the self capacitances and can be used as an approximation $\mathbf{Z}_\text{C} \approx \mathbf{Z}_\text{C}^0$ in case the self capacitances are negligible, e.g. if the operating frequency is significantly lower than the lowest self-resonance frequency of all involved coils. The matrix $\mathbf{Z}_\text{C}^0$ depends on the resistance matrix $\mathbf{R} = \text{diag}(R_1, \ldots, R_N)$ and the inductance matrix $\mathbf{M}$. The latter contains the respective self-inductances on its diagonal, i.e. $[\mathbf{M}]_{n,n} = L_n$ and its off-diagonal elements correspond to the respective mutual inductances $[\mathbf{M}]_{m,n} = M_{m,n}$ between coils m and n.

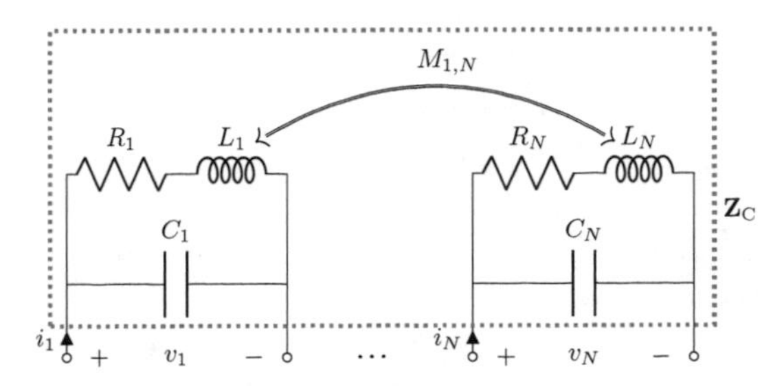

Figure 2.4: Near-field multiport circuit model for N coil antennas that are connected via impedance matrix $\mathbf{Z}_\mathrm{C}$. The model was also used by [28, 31, 52].

Chapter 3

Fundamental Theory

This chapter introduces the different types of sensor nodes used throughout this thesis and also specifies the underlying system model, which is used to simulate the MI interaction between all nodes. Additionally, it also summarizes the considered types of noise sources. The model itself is closely related to the works [28,31,97,112]. Moreover, Sec. 3.1 is supposed to be a direct extension of [28] for passive agents, as such, the structure, formulas, and nomenclature are intentionally chosen to be in line with this work. Additionally, the chapter summarizes important estimation theoretic quantities associated with this system model.

3.1 Circuit-Based Multiport Model

We consider MI sensor networks with arbitrary network constellations (also called topologies) and a variable number of synchronized sensors N. Generally, each of the sensor nodes comprises a coil antenna and additional circuitry. For each node $n = 1, \ldots, N$ we hence have to consider their six-dimensional deployment vector $\boldsymbol{\psi}_n$ (cf. Sec. 2.2). We define the full network constellation as a stacked vector containing all nodes' individual deployment vectors $\boldsymbol{\Psi}^{\text{net}} = [\boldsymbol{\psi}_1^{\text{T}}, \ldots, \boldsymbol{\psi}_N^{\text{T}}]^{\text{T}}$. We generally distinguish between three different types of sensor nodes, which all have distinct functions:

Agents are nodes which can either be active or passive, i.e. they are either connected to a dedicated current source or to a load which makes them resonant at a design frequency f^{des}. In case they are active, they may also contain matching networks e.g. to optimize the power transfer from their source into the remaining network on the design frequency. Their deployment vectors are always unknown. In the first part of this work, the underlying goal is localizing these nodes, i.e. estimating all deployment vectors $\boldsymbol{\psi}_n$ which belong to agents. In the second part of this thesis, these nodes are always passive and their deployment vectors do not need to be known. In latter case, they are also referred to as *purely passive tags*.

Anchors function as measurement infrastructure, i.e. they are connected to a measurement device and provide observations from which the deployment vectors of the agents or other information about the network shall be deduced. In the first part of this thesis, they always have a known deployment, whereas their deployment remains unknown for the second part of this work. They can also either be active or passive and contain matching networks.

Relays have a known deployment and are always considered passive i.e. they do not contain a voltage or current source. They are connected to a possibly switchable load and function as auxiliary nodes in the network. They are introduced in more detail in Cpt. 5.

The full system model is shown in Fig. 3.1 and is based on the multiport coupling model from of Sec. 2.3. It comprises $n_\mathrm{T} = 1, \ldots, N_\mathrm{T}$ agents and $n_\mathrm{R} = 1, \ldots, N_\mathrm{R}$ anchors. The explicit illustration of the relays is omitted from this representation, since their impact can be incorporated into the impedance matrix $\mathbf{Z}_\mathrm{C}$ (cf. Cpt. 5). The agents and anchors may be operated either actively or passively as schematically shown by the blue switches, which open and close in unison. In the active case, the agents have independent current sources $\mathbf{i}_\mathrm{T} = [i_{\mathrm{T},1}, \ldots, i_{\mathrm{T},N_\mathrm{T}}]^\mathrm{T}$ with internal reference resistances $R^\mathrm{ref} = 50\,\Omega$ and matching networks described by the impedance matrix $\mathbf{Z}_{\tilde{\mathrm{T}}}$. In the passive case they are generally terminated with complex impedances. Yet, in this work we only investigate passive agents that are resonant so it suffices to terminate them with capacitances $C_{\mathrm{T},1}, \ldots, C_{\mathrm{T},N_\mathrm{T}}$. The corresponding matching network of the anchors are characterized by the impedance matrix $\mathbf{Z}_{\tilde{\mathrm{R}}}$. The anchor side additionally comprises input resistances R^ref over which we measure the input currents $\mathbf{i}^\mathrm{in} = [i_1^\mathrm{in}, \ldots, i_{N_\mathrm{R}}^\mathrm{in}]^\mathrm{T}$. This input current vector $\mathbf{i}^\mathrm{in}$ comprises all observations and is the basis for our agent deployment estimation (cf. Sec. 3.4). In case the anchors are active, the input resistances R^ref simultaneously act as internal resistances for the anchor current sources $\mathbf{i}_\mathrm{R} = [i_{\mathrm{R},1}, \ldots, i_{\mathrm{R},N_\mathrm{R}}]^\mathrm{T}$. The active anchors are required to study the localization of passive agents, which only influence the input currents by changing the observable input impedances $\mathbf{Z}_\mathrm{R}^\mathrm{in}$ on the anchor side.

In order to find the input currents, we first need to fully describe all relevant impedance matrices from Fig. 3.1. These are structured in their primary (left) and

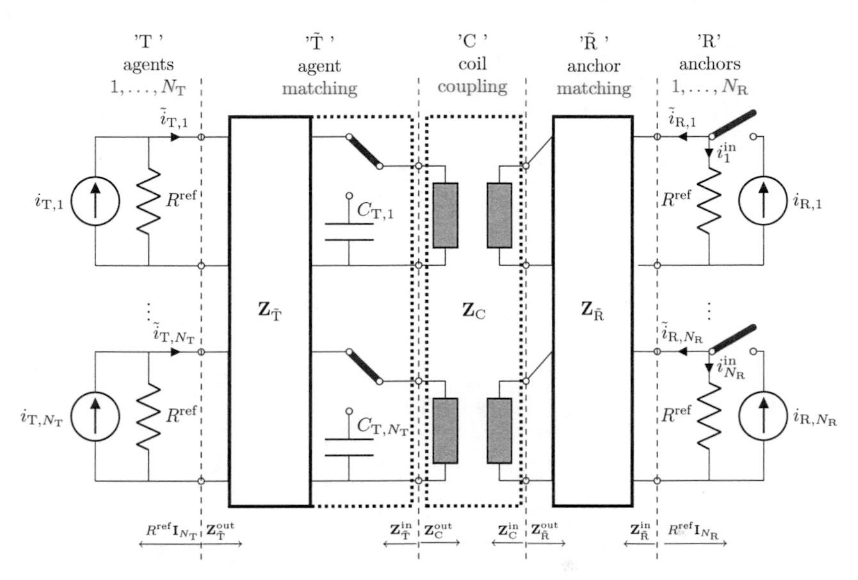

Figure 3.1: Full multiport circuit representation of agents and anchors with their corresponding matching networks $\mathbf{Z}_{\tilde{T}}, \mathbf{Z}_{\tilde{R}}$, and their coupling matrix $\mathbf{Z}_C$. The model is an extension of [28, 97].

secondary (right) sides and can be written in the form

$$\mathbf{Z}_{\tilde{T}} = \begin{bmatrix} \mathbf{Z}_{\tilde{T}:T} & \mathbf{Z}_{\tilde{T}:CT}^T \\ \mathbf{Z}_{\tilde{T}:CT} & \mathbf{Z}_{\tilde{T}:C} \end{bmatrix} \in \mathbb{C}^{2N_T \times 2N_T}, \tag{3.1}$$

$$\mathbf{Z}_C = \begin{bmatrix} \mathbf{Z}_{C:\tilde{T}} & \mathbf{Z}_{C:\tilde{T}\tilde{R}} \\ \mathbf{Z}_{C:\tilde{R}\tilde{T}} & \mathbf{Z}_{C:\tilde{R}} \end{bmatrix} \in \mathbb{C}^{(N_T+N_R) \times (N_T+N_R)}, \tag{3.2}$$

$$\mathbf{Z}_{\tilde{R}} = \begin{bmatrix} \mathbf{Z}_{\tilde{R}:C} & \mathbf{Z}_{\tilde{R}:RC}^T \\ \mathbf{Z}_{\tilde{R}:RC} & \mathbf{Z}_{\tilde{R}:R} \end{bmatrix} \in \mathbb{C}^{2N_R \times 2N_R}, \tag{3.3}$$

where the partitioning of the block matrices is performed according to the size of the corresponding primary and secondary sides, i.e. the split occurs after the N_T-th, N_T-th and N_R-th element, respectively. The individual blocks of these matrices are schematically illustrated in Fig. 3.2a. The first subscript indicates affiliation to the original blockmatrix and is separated with a colon. The second subscript indicates the side the matrix connects to, and the third subscript indicates the side the matrix connects from. If the second and third subscript are identical, i.e. the matrix connects to itself, the third subscript is dropped, e.g. $\mathbf{Z}_{C:\tilde{T}\tilde{T}} = \mathbf{Z}_{C:\tilde{T}}$. In case the self capacitances

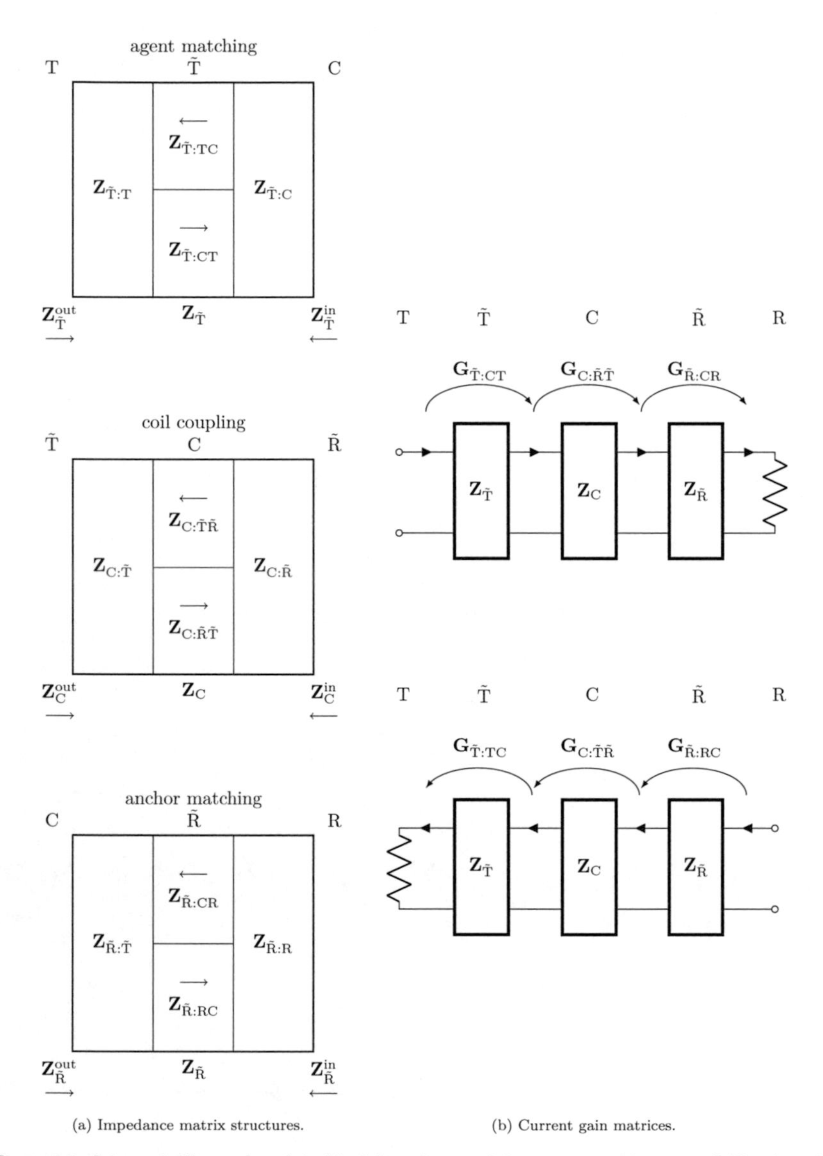

(a) Impedance matrix structures.

(b) Current gain matrices.

Figure 3.2: Schematic illustration of the block impedances of the agent matching network $\mathbf{Z}_{\tilde{T}}$, the coil coupling $\mathbf{Z}_{C}$, and the anchor matching network $\mathbf{Z}_{\tilde{R}}$. Additionally, the current gain matrices and their directions are displayed.

of the coils are negligible, the matrix $\mathbf{Z}_{\mathrm{C}:\tilde{\mathrm{T}}}$ for example describes the coupling between all matched agent coils, $\mathbf{Z}_{\mathrm{C}:\tilde{\mathrm{R}}}$ the coupling between all matched anchor coils and the so called transimpedance matrix $\mathbf{Z}_{\mathrm{C}:\tilde{\mathrm{R}}\tilde{\mathrm{T}}}$ describes the open-circuit voltages on the anchor matching side caused by the currents on the agent matching side. The associated output and input matrices of these block matrices (cf. Fig. 3.1) are given by [28, 112]

$$\mathbf{Z}_{\tilde{\mathrm{T}}}^{\mathrm{out}} = \mathbf{Z}_{\tilde{\mathrm{T}}:\mathrm{T}} - \mathbf{Z}_{\tilde{\mathrm{T}}:\mathrm{CT}}^{\mathrm{T}}(\mathbf{Z}_{\tilde{\mathrm{T}}:\mathrm{C}} + \mathbf{Z}_{\mathrm{C}}^{\mathrm{out}})^{-1}\mathbf{Z}_{\tilde{\mathrm{T}}:\mathrm{CT}}\,, \tag{3.4}$$

$$\mathbf{Z}_{\mathrm{C}}^{\mathrm{out}} = \mathbf{Z}_{\mathrm{C}:\tilde{\mathrm{T}}} - \mathbf{Z}_{\mathrm{C}:\tilde{\mathrm{R}}\tilde{\mathrm{T}}}^{\mathrm{T}}(\mathbf{Z}_{\mathrm{C}:\tilde{\mathrm{R}}} + \mathbf{Z}_{\tilde{\mathrm{R}}}^{\mathrm{out}})^{-1}\mathbf{Z}_{\mathrm{C}:\tilde{\mathrm{R}}\tilde{\mathrm{T}}}\,, \tag{3.5}$$

$$\mathbf{Z}_{\tilde{\mathrm{R}}}^{\mathrm{out}} = \mathbf{Z}_{\tilde{\mathrm{R}}:\mathrm{C}} - \mathbf{Z}_{\tilde{\mathrm{R}}:\mathrm{CR}}^{\mathrm{T}}(\mathbf{Z}_{\tilde{\mathrm{R}}:\mathrm{R}} + R^{\mathrm{ref}}\mathbf{I}_{N_\mathrm{R}})^{-1}\mathbf{Z}_{\tilde{\mathrm{R}}:\mathrm{CR}}\,, \tag{3.6}$$

$$\mathbf{Z}_{\tilde{\mathrm{T}}}^{\mathrm{in}} = \mathbf{Z}_{\tilde{\mathrm{T}}:\mathrm{C}} - \mathbf{Z}_{\tilde{\mathrm{T}}:\mathrm{CT}}(\mathbf{Z}_{\tilde{\mathrm{T}}:\mathrm{T}} + R^{\mathrm{ref}}\mathbf{I}_{N_\mathrm{T}})^{-1}\mathbf{Z}_{\tilde{\mathrm{T}}:\mathrm{CT}}^{\mathrm{T}}\,, \tag{3.7}$$

$$\mathbf{Z}_{\mathrm{C}}^{\mathrm{in}} = \mathbf{Z}_{\mathrm{C}:\tilde{\mathrm{R}}} - \mathbf{Z}_{\mathrm{C}:\tilde{\mathrm{R}}\tilde{\mathrm{T}}}(\mathbf{Z}_{\mathrm{C}:\tilde{\mathrm{T}}} + \mathbf{Z}_{\tilde{\mathrm{T}}}^{\mathrm{in}})^{-1}\mathbf{Z}_{\mathrm{C}:\tilde{\mathrm{R}}\tilde{\mathrm{T}}}^{\mathrm{T}}\,, \tag{3.8}$$

$$\mathbf{Z}_{\tilde{\mathrm{R}}}^{\mathrm{in}} = \mathbf{Z}_{\tilde{\mathrm{R}}:\mathrm{R}} - \mathbf{Z}_{\tilde{\mathrm{R}}:\mathrm{CR}}(\mathbf{Z}_{\tilde{\mathrm{R}}:\mathrm{C}} + \mathbf{Z}_{\mathrm{C}}^{\mathrm{in}})^{-1}\mathbf{Z}_{\tilde{\mathrm{R}}:\mathrm{CR}}^{\mathrm{T}}\,. \tag{3.9}$$

Note that in case of passive agents we have $\mathbf{Z}_{\tilde{\mathrm{T}}}^{\mathrm{in}} = \mathbf{Z}_{\mathrm{T}}^{\mathrm{load}} = \frac{1}{j\omega}\,\mathrm{diag}(C_{\mathrm{T},1}^{\mathrm{load}}, \ldots, C_{\mathrm{T},N_\mathrm{T}}^{\mathrm{load}})^{-1}$ containing the agents' load capacitances and $\mathbf{Z}_{\tilde{\mathrm{T}}}^{\mathrm{out}} = Z_\infty \mathbf{I}_{N_\mathrm{T}}$, where $Z_\infty \to \infty$ represents the high resistance of an open circuit.

Moreover, the parts of the source currents which end up being fed into corresponding matching networks are given by

$$\tilde{\mathbf{i}}_\mathrm{T} = (R^{\mathrm{ref}}\mathbf{I}_{N_\mathrm{T}} + \mathbf{Z}_{\tilde{\mathrm{T}}}^{\mathrm{out}})^{-1}R^{\mathrm{ref}}\mathbf{i}_\mathrm{T}\,, \tag{3.10}$$

$$\tilde{\mathbf{i}}_\mathrm{R} = (R^{\mathrm{ref}}\mathbf{I}_{N_\mathrm{R}} + \mathbf{Z}_{\tilde{\mathrm{R}}}^{\mathrm{in}})^{-1}R^{\mathrm{ref}}\mathbf{i}_\mathrm{R}\,, \tag{3.11}$$

and we obtain the active power of the n_T-th agent source and the n_R-th anchor source into the multiport system as [28, Eq. 3.12]

$$P_{\mathrm{T},n_\mathrm{T}} = \mathrm{Re}\left[\mathbf{Z}_{\tilde{\mathrm{T}}}^{\mathrm{out}}\tilde{\mathbf{i}}_\mathrm{T}\tilde{\mathbf{i}}_\mathrm{T}^{\mathrm{H}}\right]_{n_\mathrm{T},n_\mathrm{T}}\,, \tag{3.12}$$

$$P_{\mathrm{R},n_\mathrm{R}} = \mathrm{Re}\left[\mathbf{Z}_{\tilde{\mathrm{R}}}^{\mathrm{in}}\tilde{\mathbf{i}}_\mathrm{R}\tilde{\mathbf{i}}_\mathrm{R}^{\mathrm{H}}\right]_{n_\mathrm{R},n_\mathrm{R}}\,. \tag{3.13}$$

We next look at the current gain matrices through each multiport, which relate the currents into a multiport to the currents out of a multiport and are found by considering one side as partially terminated [28]. A schematic illustration of these current gain matrices and their directions is provided in Fig. 3.2b. Formally, they are given by

$$\mathbf{G}_{\tilde{\mathrm{T}}:\mathrm{CT}} = (\mathbf{Z}_{\mathrm{C}}^{\mathrm{out}} + \mathbf{Z}_{\tilde{\mathrm{T}}:\mathrm{C}})^{-1}\mathbf{Z}_{\tilde{\mathrm{T}}:\mathrm{CT}}\,, \qquad \mathbf{G}_{\tilde{\mathrm{T}}:\mathrm{TC}} = (R^{\mathrm{ref}}\mathbf{I}_{N_{\mathrm{T}}} + \mathbf{Z}_{\tilde{\mathrm{T}}:\mathrm{T}})^{-1}\mathbf{Z}_{\tilde{\mathrm{T}}:\mathrm{CT}}^{\mathrm{T}}\,, \tag{3.14}$$

$$\mathbf{G}_{\mathrm{C}:\tilde{\mathrm{R}}\tilde{\mathrm{T}}} = (\mathbf{Z}_{\tilde{\mathrm{R}}}^{\mathrm{out}} + \mathbf{Z}_{\mathrm{C}:\tilde{\mathrm{R}}})^{-1}\mathbf{Z}_{\mathrm{C}:\tilde{\mathrm{R}}\tilde{\mathrm{T}}}\,, \qquad \mathbf{G}_{\mathrm{C}:\tilde{\mathrm{T}}\tilde{\mathrm{R}}} = (\mathbf{Z}_{\tilde{\mathrm{T}}}^{\mathrm{in}} + \mathbf{Z}_{\mathrm{C}:\tilde{\mathrm{T}}})^{-1}\mathbf{Z}_{\mathrm{C}:\tilde{\mathrm{R}}\tilde{\mathrm{T}}}^{\mathrm{T}}\,, \tag{3.15}$$

$$\mathbf{G}_{\tilde{\mathrm{R}}:\mathrm{RC}} = (R^{\mathrm{ref}}\mathbf{I}_{N_{\mathrm{R}}} + \mathbf{Z}_{\tilde{\mathrm{R}}:\mathrm{R}})^{-1}\mathbf{Z}_{\tilde{\mathrm{R}}:\mathrm{RC}}\,, \qquad \mathbf{G}_{\tilde{\mathrm{R}}:\mathrm{CR}} = (\mathbf{Z}_{\mathrm{C}}^{\mathrm{in}} + \mathbf{Z}_{\tilde{\mathrm{R}}:\mathrm{C}})^{-1}\mathbf{Z}_{\tilde{\mathrm{R}}:\mathrm{RC}}^{\mathrm{T}}\,, \tag{3.16}$$

with the total current gain matrix over all concatenated multiports being the product of the individual constituents, e.g. from agents to anchors

$$\mathbf{G}^{\mathrm{active}} = \mathbf{G}_{\tilde{\mathrm{R}}:\mathrm{RC}}\,\mathbf{G}_{\mathrm{C}:\tilde{\mathrm{R}}\tilde{\mathrm{T}}}\,\mathbf{G}_{\tilde{\mathrm{T}}:\mathrm{CT}}\,. \tag{3.17}$$

For consistency reasons, we also define an artificial passive current gain matrix that connects the anchor currents into the network $\mathbf{i}_{\tilde{\mathrm{R}}}^{\mathrm{in}}$ to the observed input currents $\mathbf{i}^{\mathrm{in}}$

$$\mathbf{G}^{\mathrm{passive}} = (R^{\mathrm{ref}}\mathbf{I}_{N_{\mathrm{R}}} + \mathbf{Z}_{\tilde{\mathrm{R}}}^{\mathrm{in}})^{-1}\,\mathbf{Z}_{\tilde{\mathrm{R}}}^{\mathrm{in}}\,(R^{\mathrm{ref}})^{-1}\,(R^{\mathrm{ref}}\mathbf{I}_{N_{\mathrm{R}}} + \mathbf{Z}_{\tilde{\mathrm{R}}}^{\mathrm{in}})\,. \tag{3.18}$$

With these quantities, the observation input current vectors follow as

$$\mathbf{i}_{\mathrm{T}}^{\mathrm{in}} = \mathbf{G}^{\mathrm{active}}\,\tilde{\mathbf{i}}_{\mathrm{T}}\,, \quad \text{(active agent, passive anchor)}\,, \tag{3.19}$$

$$\mathbf{i}_{\mathrm{R}}^{\mathrm{in}} = \mathbf{G}^{\mathrm{passive}}\,\tilde{\mathbf{i}}_{\mathrm{R}}\,, \quad \text{(passive agent, active anchor)}\,, \tag{3.20}$$

with $\mathbf{i}_{\mathrm{T}}^{\mathrm{in}}$ and $\mathbf{i}_{\mathrm{R}}^{\mathrm{in}}$ being the input current vectors in case active or passive agents are used, respectively. The quantity $\mathbf{i}^{\mathrm{in}}$ will be used to describe either of the two. Also note that there is a significant input current vector $\mathbf{i}_{\mathrm{R}}^{\mathrm{in}}$ even if no agents are present in the network due to the currents from the anchor sources as well as the inter-anchor coupling. Only the small difference of $\mathbf{i}_{\mathrm{R}}^{\mathrm{in}}$ that is caused by the presence of the agents actually conveys information on the agent deployments.

3.2 Noise Model

In reality, the input currents will also be affected by additional perturbations e.g. due to thermal noise and imperfections of **L**ow-**N**oise **A**mplifiers (LNAs). We model these perturbations by additive circularly-symmetric zero-mean Gaussian noise currents, i.e. [97]

$$\mathbf{i}^{\mathrm{meas}} = \mathbf{i}^{\mathrm{in}} + \mathbf{i}^{w} \qquad \mathbf{i}^{w} \sim \mathcal{CN}(0, \mathbf{K})\,, \tag{3.21}$$

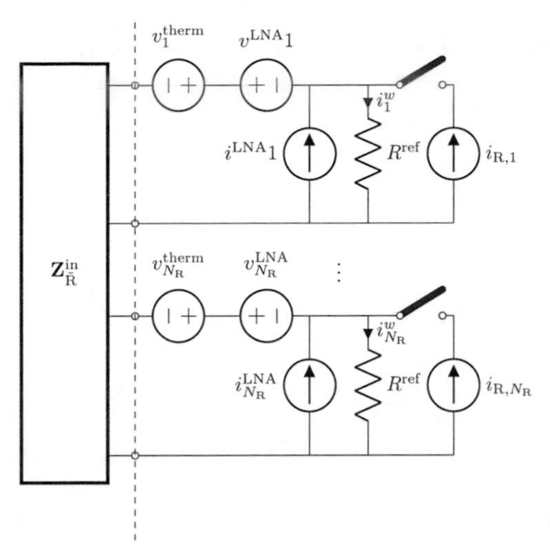

Figure 3.3: Anchor side with considered noise sources due to Johnson–Nyquist noise and LNA imperfections. The model is adapted from [28, 97].

The noisy currents at each anchor are a result of a voltage source for the thermal noise as well as a possibly correlated voltage-current source pair, as illustrated in Fig. 3.3. The overall noise covariance matrix for the currents is given by [28, Sec. 3.2]

$$\mathbf{K} = \mathbf{Y}_\text{R}(\mathbf{\Sigma}^{\text{therm}} + \mathbf{\Sigma}^{\text{LNA}})\mathbf{Y}_\text{R}^\text{H} \tag{3.22}$$

$$\mathbf{Y}_\text{R} = (R^{\text{ref}}\mathbf{I}_{N_\text{R}} + \mathbf{Z}_{\bar{\text{R}}}^{\text{in}})^{-1} \tag{3.23}$$

$$\mathbf{\Sigma}^{\text{therm}} = 4k_\text{B}TB\text{Re}(\mathbf{Z}_{\bar{\text{R}}}^{\text{in}}) \tag{3.24}$$

$$\mathbf{\Sigma}^{\text{LNA}} = (\sigma^{\text{LNA}})^2\left((R^{\text{LNA}})^2\mathbf{I}_{N_\text{R}} + \mathbf{Z}_{\bar{\text{R}}}^{\text{in}}(\mathbf{Z}_{\bar{\text{R}}}^{\text{in}})^\text{H} - 2R^{\text{LNA}}\text{Re}((\rho^{\text{LNA}})^*\mathbf{Z}_{\bar{\text{R}}}^{\text{in}})\right), \tag{3.25}$$

with $\mathbf{Y}_\text{R}$ as the serial admittance of the anchor circuit. For the covariance matrix $\mathbf{\Sigma}^{\text{therm}}$ of the thermal noise, only the real part of the anchor's input impedance $\mathbf{Z}_{\bar{\text{R}}}^{\text{in}}$ is relevant [113]. Additionally, the thermal noise depends on the Boltzmann constant k_B, the temperature T and the bandwidth B. The noisy currents caused by all LNA source pairs have the covariance matrix $\mathbf{\Sigma}^{\text{LNA}}$, which is defined via noise current variance $(\sigma^{\text{LNA}})^2$, noise resistance R^{LNA}, and the correlation coefficient ρ^{LNA} with $\left\|\rho^{\text{LNA}}\right\| \leq 1$. The LNAs are also assumed to be statistically independent when the anchor coils and anchor matching networks are uncoupled.

3.3 Matching, Equivalent Model, and Further Assumptions

3.3.1 Matching Networks

The matching networks can be chosen according to the needs of the system. Typical choices include lossless power matching on both sides or noise matching on the receiver side, e.g. to obtain a high **S**ignal-to-**N**oise **R**atio (SNR) for wireless communications [97,112]. For lossless matching networks, the corresponding impedance matrices need to be symmetric and purely imaginary. At a chosen design frequency f^{des}, ideal multiport power matching requires

$$\mathbf{Z}_{\tilde{\mathrm{T}}}^{\mathrm{out}} = R^{\mathrm{ref}}\mathbf{I}_{N_{\mathrm{T}}}\,, \qquad \mathbf{Z}_{\tilde{\mathrm{T}}}^{\mathrm{in}} = (\mathbf{Z}_{\mathrm{C}}^{\mathrm{out}})^*\,, \qquad \text{(ideal agent power matching)}\,, \tag{3.26}$$

$$\mathbf{Z}_{\tilde{\mathrm{R}}}^{\mathrm{in}} = R^{\mathrm{ref}}\mathbf{I}_{N_{\mathrm{R}}}\,, \qquad \mathbf{Z}_{\tilde{\mathrm{R}}}^{\mathrm{out}} = (\mathbf{Z}_{\mathrm{C}}^{\mathrm{in}})^*\,, \qquad \text{(ideal anchor power matching)}\,. \tag{3.27}$$

Individually, this can be realized by using matching networks with the following impedance matrices [97]

$$\mathbf{Z}_{\tilde{\mathrm{T}}} = \begin{bmatrix} \mathbf{0}_{N_{\mathrm{T}}} & \pm j\sqrt{R^{\mathrm{ref}}}\mathrm{Re}(\mathbf{Z}_{\mathrm{C}}^{\mathrm{out}})^{\frac{1}{2}} \\ \pm j\sqrt{R^{\mathrm{ref}}}\mathrm{Re}(\mathbf{Z}_{\mathrm{C}}^{\mathrm{out}})^{\frac{1}{2}} & -j\mathrm{Im}(\mathbf{Z}_{\mathrm{C}}^{\mathrm{out}}) \end{bmatrix}, \tag{3.28}$$

$$\mathbf{Z}_{\tilde{\mathrm{R}}} = \begin{bmatrix} -j\mathrm{Im}(\mathbf{Z}_{\mathrm{C}}^{\mathrm{in}}) & \pm j\sqrt{R^{\mathrm{ref}}}\mathrm{Re}(\mathbf{Z}_{\mathrm{C}}^{\mathrm{in}})^{\frac{1}{2}} \\ \pm j\sqrt{R^{\mathrm{ref}}}\mathrm{Re}(\mathbf{Z}_{\mathrm{C}}^{\mathrm{in}})^{\frac{1}{2}} & \mathbf{0}_{N_{\mathrm{R}}} \end{bmatrix}. \tag{3.29}$$

However, due to the coupling of the agents and anchors the matching networks affect one another, which complicates a joint matching of both sides. Although iterative matching approaches are being investigated, to the best of our knowledge there is no analytic closed-form solution that matches both sides jointly for the **M**ultiple **I**nput **M**ultiple **O**utput (MIMO) case. Moreover, note that actual multiport matching networks can be hard to realize in practice as they require a vast amount of lumped elements with a low component tolerance, e.g. $2N^2 + N$ lumped elements to realize a Π-structured multiport matching network for N coils [114]. Further, for dynamic network constellations, the matching networks require to be adaptive to satisfy either (3.26) or (3.27).

A different low-complexity approach uses individual fixed two-port power matching networks for each active coil. That is, the overall matching network consists of individual and lossless two-port networks that each power match the reference resistance to the corresponding coil antenna impedance in the absence of other coils. For the n_{T}-th agent or the n_{R}-th anchor coil such a two-port network can e.g. have the impedance matrix (cf. (3.28) and (3.29))

$$\mathbf{Z}_{\tilde{\mathrm{T}},n_{\mathrm{T}}}^{\text{two-port}} = \begin{bmatrix} 0 & \pm j\sqrt{R^{\mathrm{ref}}}\mathrm{Re}(Z_{\mathrm{T},n_{\mathrm{T}}})^{\frac{1}{2}} \\ \pm j\sqrt{R^{\mathrm{ref}}}\mathrm{Re}(Z_{\mathrm{T},n_{\mathrm{T}}})^{\frac{1}{2}} & -j\mathrm{Im}(Z_{\mathrm{T},n_{\mathrm{T}}}) \end{bmatrix}, \tag{3.30}$$

$$\mathbf{Z}_{\tilde{\mathrm{R}},n_{\mathrm{R}}}^{\text{two-port}} = \begin{bmatrix} -j\mathrm{Im}(Z_{\mathrm{R},n_{\mathrm{R}}}) & \pm j\sqrt{R^{\mathrm{ref}}}\mathrm{Re}(Z_{\mathrm{R},n_{\mathrm{R}}})^{\frac{1}{2}} \\ \pm j\sqrt{R^{\mathrm{ref}}}\mathrm{Re}(Z_{\mathrm{R},n_{\mathrm{R}}})^{\frac{1}{2}} & 0 \end{bmatrix}, \tag{3.31}$$

where $Z_{\mathrm{T},n_{\mathrm{T}}}$ and $Z_{\mathrm{R},n_{\mathrm{R}}}$ are the impedances of the corresponding agent and anchor coil antennas including their self capacitances (cf. (2.10)). For strongly-coupled coils, i.e. when the impedances of $\mathbf{Z}_{\mathrm{C}}^{\mathrm{out}}$ and $\mathbf{Z}_{\mathrm{C}}^{\mathrm{in}}$ are strongly detuned from their original coil antenna impedances, this approach clearly leads to a mismatch of the matching networks and hence violates (3.26) and (3.27). Such a mismatch results in a reduced power transfer into the network and may be detrimental for communication and localization. Using adaptive matching networks is one intricate and possibly costly solution to mitigate this issue. Alternatively, it is also possible to employ other current sources which provide a constant current (and hence increasing power) into the network, regardless of the impedance detuning. Yet, neither of these alternatives may be practical for low-complexity applications which have to operate on a small scale.

Lastly, for passive agents we generally consider no additional matching networks and the agent coils are always loaded with the corresponding load capacitors. For this case the impedance matrix of the agent matching network can be mathematically represented by $\mathbf{Z}_{\tilde{\mathrm{T}}} = \begin{bmatrix} Z_{\infty}\mathbf{I}_{N_{\mathrm{T}}} & \mathbf{0}_{N_{\mathrm{T}}} \\ \mathbf{0}_{N_{\mathrm{T}}} & \mathbf{Z}_{\tilde{\mathrm{T}}}^{\mathrm{load}} \end{bmatrix}$.

3.3.2 Equivalent Channel Gain Description

Instead of using currents, communication theoretic works often rely on a normalized signal model to provide mathematical convenience in various circumstances. The MIMO representations of (3.19) and (3.20) can be transformed to such models of normalized signal vectors $\mathbf{x}$ and observation vectors $\mathbf{y}$ by applying the following substitutions [28,97]

$$\mathbf{x}_\mathrm{T} = \mathrm{Re}(\mathbf{Z}^\mathrm{out}_{\tilde{\mathrm{T}}})^{\frac{1}{2}}\,\tilde{\mathbf{i}}_\mathrm{T}\,, \qquad \mathbf{x}_\mathrm{R} = \mathrm{Re}(\mathbf{Z}^\mathrm{in}_{\tilde{\mathrm{R}}})^{\frac{1}{2}}\,\tilde{\mathbf{i}}_\mathrm{R}\,, \tag{3.32}$$

$$\mathbf{H}_\mathrm{T} = (R^\mathrm{ref})^{\frac{1}{2}}\mathbf{G}^\mathrm{active}\mathrm{Re}(\mathbf{Z}^\mathrm{out}_{\tilde{\mathrm{T}}})^{-\frac{1}{2}}\,, \qquad \mathbf{H}_\mathrm{R} = (R^\mathrm{ref})^{\frac{1}{2}}\mathbf{G}^\mathrm{passive}\,\mathrm{Re}(\mathbf{Z}^\mathrm{in}_{\tilde{\mathrm{R}}})^{-\frac{1}{2}}\,, \tag{3.33}$$

$$\mathbf{y} = (R^\mathrm{ref})^{\frac{1}{2}}\mathbf{i}^\mathrm{in}\,, \qquad \mathbf{y}^\mathrm{meas} = (R^\mathrm{ref})^{\frac{1}{2}}\mathbf{i}^\mathrm{meas}\,, \tag{3.34}$$

$$\mathbf{w} = (R^\mathrm{ref})^{\frac{1}{2}}\mathbf{i}^w\,, \tag{3.35}$$

which, analogously to the input current model (cf. (3.19) and (3.20)), leads to

$$\mathbf{y}^\mathrm{meas}_\mathrm{T} = \mathbf{y}_\mathrm{T} + \mathbf{w} = \mathbf{H}_\mathrm{T}\mathbf{x}_\mathrm{T} + \mathbf{w}\,, \quad \text{(active agent, passive anchor)}\,, \tag{3.36}$$

$$\mathbf{y}^\mathrm{meas}_\mathrm{R} = \mathbf{y}_\mathrm{R} + \mathbf{w} = \mathbf{H}_\mathrm{R}\mathbf{x}_\mathrm{R} + \mathbf{w} \quad \text{(passive agent, active anchor)}\,. \tag{3.37}$$

With these formulations, it is convenient to describe the **P**ower **T**ransfer **E**fficiency (PTE) η, which is the corresponding ratio of active receive sum-power $\|\mathbf{y}\|^2$ (receive power) to active transmit sum-power $\|\mathbf{x}\|^2$ (transmit power) without additional noise, i.e. [28, Eq. 3.27]

$$\eta_\mathrm{T} = \frac{\|\mathbf{y}_\mathrm{T}\|^2}{\|\mathbf{x}_\mathrm{T}\|^2} = \frac{\|\mathbf{H}_\mathrm{T}\mathbf{x}_\mathrm{T}\|^2}{\|\mathbf{x}_\mathrm{T}\|^2} \overset{SIMO}{=} \frac{\|\mathbf{h}_\mathrm{T}\|^2\,|x_\mathrm{T}|^2}{|x_\mathrm{T}|^2} = \|\mathbf{h}_\mathrm{T}\|^2 \overset{SISO}{=} |h_\mathrm{T}|^2\,, \tag{3.38}$$

$$\eta_\mathrm{R} = \frac{\|\mathbf{y}_\mathrm{R}\|^2}{\|\mathbf{x}_\mathrm{R}\|^2} = \frac{\|\mathbf{H}_\mathrm{R}\mathbf{x}_\mathrm{R}\|^2}{\|\mathbf{x}_\mathrm{R}\|^2} \overset{SISO}{=} |h_\mathrm{R}|^2\,, \tag{3.39}$$

where the simpler formulations for the **S**ingle **I**nput **M**ultiple **O**utput (SIMO) and **S**ingle **I**nput **S**ingle **O**utput (SISO) case are independent of the transmit power. For the passive case, all anchors act simultaneously as transmitter and receiver, so there is no SIMO case.

3.3.3 Unilateral Assumption for Weakly-Coupled Links

The wireless MI coupling of our model is in general reciprocal, i.e. $\mathbf{Z}_{\mathrm{C}:\tilde{\mathrm{T}}\tilde{\mathrm{R}}} = \mathbf{Z}^\mathrm{T}_{\mathrm{C}:\tilde{\mathrm{T}}\tilde{\mathrm{R}}}$. For weakly-coupled agent and anchor sides combined with a unidirectional communication, the resulting reverse channel (i.e. from the passive to the active side) can often be neglected. This is referred to as the unilateral assumption [97]. It is commonly applied in radio communications and also holds well for the magnetic near field if the involved coils are separated by a large distance, which is e.g. the case for sparse networks. We will rely on this assumption multiple times throughout this thesis as it simplifies the matching procedure and makes analytic investigations more convenient. For the active agent case, it is mathematically represented by $\mathbf{Z}_{\mathrm{C:\tilde{T}\tilde{R}}} = \mathbf{0}_{N_\mathrm{T} \times N_\mathrm{R}}$, however $\mathbf{Z}_{\mathrm{C:\tilde{R}\tilde{T}}} \neq \mathbf{0}_{N_\mathrm{R} \times N_\mathrm{T}}$, which decouples the impedances on both sides of the multiport $\mathbf{Z}_\mathrm{C}$, i.e. $\mathbf{Z}_\mathrm{C}^\mathrm{out} = \mathbf{Z}_{\mathrm{C:\tilde{T}}}$ and $\mathbf{Z}_\mathrm{C}^\mathrm{in} = \mathbf{Z}_{\mathrm{C:\tilde{R}}}$. If we combine this assumption with an ideal power matching on both sides, (3.4) to (3.9) can be simplified as

$$\mathbf{Z}_{\tilde{\mathrm{T}}}^\mathrm{out} = R^\mathrm{ref}\mathbf{I}_{N_\mathrm{T}}\,, \qquad \mathbf{Z}_{\tilde{\mathrm{T}}}^\mathrm{in} = (\mathbf{Z}_\mathrm{C}^\mathrm{out})^*\,, \tag{3.40}$$

$$\mathbf{Z}_\mathrm{C}^\mathrm{out} = \mathbf{Z}_{\mathrm{C:\tilde{T}}}\,, \qquad \mathbf{Z}_\mathrm{C}^\mathrm{in} = \mathbf{Z}_{\mathrm{C:\tilde{R}}}\,, \tag{3.41}$$

$$\mathbf{Z}_{\tilde{\mathrm{R}}}^\mathrm{out} = (\mathbf{Z}_\mathrm{C}^\mathrm{in})^*\,, \qquad \mathbf{Z}_{\tilde{\mathrm{R}}}^\mathrm{in} = R^\mathrm{ref}\mathbf{I}_{N_\mathrm{R}}\,, \tag{3.42}$$

which yields (cf. (3.10) and (3.17))

$$\mathbf{G}^\mathrm{active} = \mathbf{H}_\mathrm{T} = \frac{1}{2}\mathrm{Re}(\mathbf{Z}_{\mathrm{C:\tilde{R}}})^{-\frac{1}{2}}\mathbf{Z}_{\mathrm{C:\tilde{R}\tilde{T}}}\mathrm{Re}(\mathbf{Z}_{\mathrm{C:\tilde{T}}})^{-\frac{1}{2}}\,, \tag{3.43}$$

$$\mathbf{i}_\mathrm{T}^\mathrm{in} = \frac{1}{2}\mathbf{G}^\mathrm{active}\mathbf{i}_\mathrm{T}\,. \tag{3.44}$$

For a SISO link of an active agent m and a passive anchor n with coil resistances R_m, R_n, mutual inductance $M_{m,n}$, and negligible self capacitances, (3.43) can be further simplified such that the scalar current gain g^active and the scalar channel gain h_T are identical and expressed by

$$g^\mathrm{active} = h_\mathrm{T} = \frac{j\omega M_{m,n}}{\sqrt{4R_mR_n}}\,. \tag{3.45}$$

While the unilateral assumption is easily applied to the unidirectional active agent case, it contradicts the bidirectional passive agent case. Moreover, even if we assume the passive agents and active anchors to be weakly coupled such that the reverse channel (from agent to anchors) does not meaningfully affect the anchors, then no information about the agents may be derived from the input currents on the anchor side. The same also holds for adaptive matching, which for ideal power matching networks on the anchor side satisfies (3.42) and hence nullifies the agent impact, leading to (cf. (3.11) and (3.18))

$$\mathbf{G}^\mathrm{passive} = \mathbf{H}_\mathrm{R} = \mathbf{I}_{N_\mathrm{T}}\,, \tag{3.46}$$

$$\mathbf{i}_\mathrm{R}^\mathrm{in} = \mathbf{G}^\mathrm{passive}\frac{1}{2}\mathbf{i}_\mathrm{R} = \frac{1}{2}\mathbf{i}_\mathrm{R}\,. \tag{3.47}$$

3.4 Parameter Estimation and Cramér-Rao Lower Bound

For the general case without the unilateral assumption, the observed input currents always depend on all unknown agent deployments via mutual inductances between the coils. The localization can be formulated in terms of a parameter estimation problem of our model, where we define the agent constellation as stacked vector $\boldsymbol{\Psi} = [\boldsymbol{\psi}_1^{\mathrm{T}}, \dots, \boldsymbol{\psi}_{N_{\mathrm{T}}}^{\mathrm{T}}]^{\mathrm{T}}$ which contains all individual agent deployment vectors. The joint ML estimate follows as

$$\hat{\boldsymbol{\Psi}}^{\mathrm{ML}} = \arg\min_{\boldsymbol{\Psi}} \left\{ \left(\mathbf{i}^{\mathrm{meas}} - \mathbf{i}^{\mathrm{in}}(\boldsymbol{\Psi})\right)^{\mathrm{H}} \mathbf{K}^{-1} \left(\mathbf{i}^{\mathrm{meas}} - \mathbf{i}^{\mathrm{in}}(\boldsymbol{\Psi})\right) \right\} . \tag{3.48}$$

However, to the best of our knowledge there is no analytic closed-form solution for the estimator (3.48), due to the intricate relationship between deployment variables and currents. Yet, for the Gaussian estimation problem at hand, the ML estimator is known to be asymptotically efficient, i.e. it approaches the CRLB as the number of independent observations goes to infinity [115]. We will hence analyze different setups based on the corresponding **P**osition **E**rror **B**ound (PEB), which is the CRLB on the position RMSE error and whose calculation requires the **F**isher **I**nformation **M**atrix (FIM) $\boldsymbol{\mathcal{I}}$. For a Gaussian **P**robability **D**ensity **F**unction (PDF) with real parameters and complex observations, the scalar elements of the FIM are given by a well-known expression [115, Eq. 15.52]

$$[\boldsymbol{\mathcal{I}}(\boldsymbol{\Psi})]_{i,j} = \mathrm{tr}\left(\mathbf{K}^{-1}\frac{\partial \mathbf{K}}{\partial [\boldsymbol{\Psi}]_i}\mathbf{K}^{-1}\frac{\partial \mathbf{K}}{\partial [\boldsymbol{\Psi}]_j}\right) + 2\mathrm{Re}\left(\frac{\partial (\mathbf{i}^{\mathrm{in}})^{\mathrm{H}}}{\partial [\boldsymbol{\Psi}]_i}\mathbf{K}^{-1}\frac{\partial \mathbf{i}^{\mathrm{in}}}{\partial [\boldsymbol{\Psi}]_j}\right) . \tag{3.49}$$

In many scenarios, e.g. when the agent is active and the input impedance matrix of the anchor matching $\mathbf{Z}_{\mathrm{R}}^{\mathrm{in}}$ does not vary significantly for deployment changes of the agents, the first summand can be neglected. For such scenarios we can approximate the full FIM as

$$\boldsymbol{\mathcal{I}}(\boldsymbol{\Psi}) \approx 2\mathrm{Re}\left(\mathbf{J}_{\mathbf{i}^{\mathrm{in}}}^{\mathrm{H}}\, \mathbf{K}^{-1}\, \mathbf{J}_{\mathbf{i}^{\mathrm{in}}}\right) , \tag{3.50}$$

where $\mathbf{J}_{\mathbf{i}^{\mathrm{in}}}$ is a complex-valued Jacobian with its j-th column given by $[\mathbf{J}_{\mathbf{i}^{\mathrm{in}}}]_{:,j} = \frac{\partial \mathbf{i}^{\mathrm{in}}}{\partial [\boldsymbol{\Psi}]_j}$. In plain words, the accuracy with which we can estimate the deployment variables depends on how severely changes of these variables impact the observed input currents

$\mathbf{i}^{\text{in}}$ relative to the intensity of the additional noise $\mathbf{K}^{-1}$. Moreover, we call the Fisher information of any element of the position vector $\mathbf{p}_m$ of an agent m *spatial information* and Fisher information of any element of the orientation vector $\boldsymbol{\phi}_m$ *orientational* information. Additionally, considering a unit vector $\mathbf{q}$, we call $\mathbf{q}^{\text{T}}\left[\boldsymbol{\mathcal{I}}(\boldsymbol{\Psi})\right]_{1:3,1:3}\mathbf{q}$ *directional* (spatial) information of the first agent in the direction $\mathbf{q}$, with analogous definition for the other agents.

Furthermore, when combining $k = 1, \ldots, N_k$ statistically independent measurements for a static network constellation, which may occur when combining measurements from multiple distant anchors or when collecting measurements for the same network constellation at different time instances, the overall FIM is equal to the sum of the individual FIMs $\boldsymbol{\mathcal{I}}_k$ (additivity of independent Fisher information [116, 117]):

$$\boldsymbol{\mathcal{I}}(\boldsymbol{\Psi}) = \sum_{k=1}^{N_K} 2\text{Re}\left(\mathbf{J}_{\mathbf{i}^{\text{in}},k}^{\text{H}}\,\mathbf{K}_k^{-1}\,\mathbf{J}_{\mathbf{i}^{\text{in}},k}\right)\,, \tag{3.51}$$

where $\mathbf{J}_{\mathbf{i}^{\text{in}},k}$ is the complex-valued Jacobian matrix associated with the k-th measurement.

Considering only the first agent, we find the PEB of any unbiased position estimator $\hat{\mathbf{p}}_1 = [\hat{p}_{1,x}, \hat{p}_{1,y}, \hat{p}_{1,z}]^{\text{T}}$ via [118]

$$\text{PEB}_1(\boldsymbol{\Psi}) = \sqrt{\text{tr}\left[\left[\boldsymbol{\mathcal{I}}^{-1}(\boldsymbol{\Psi})\right]_{1:3,1:3}\right]} \tag{3.52}$$

$$\leq \sqrt{\mathbb{E}\{||\hat{\mathbf{p}}_1 - \mathbf{p}_1||^2\}}. \tag{3.53}$$

Apart from all known and fixed system quantities, the calculation of the PEB requires the respective Jacobians. The corresponding derivatives are simply found by iteratively applying basic matrix derivation rules (cf. Appendix A). It is shown that the evaluation of (3.49) is ultimately based on all mutual inductances $M_{m,n}$ and the derivatives $\frac{\partial M_{m,n}}{\partial[\boldsymbol{\Psi}]_j}$ thereof. We can obtain them by either modeling the coils as magnetic dipoles, which leads to the well known approximation (2.3) of their mutual inductances whose derivatives are found in [26]. This approximation is accurate as long as the distance between any two coils is much larger than their radii. For closer coil proximities, the double integral Neumann formula (2.2) can be used but generally requires a numerical integration. Up to this numerical integration, the derivatives of the Neumann formula for solenoid coils are also stated in Appendix A.

Being able to calculate the FIM and the scalar PEBs of any agent for a given network constellation, we also want to analyze how the position RMSE error is bounded in any unit direction $\mathbf{q}$. Analogous to [118], we call this bound **D**irectional **P**osition **E**rror **B**ound (DPEB) and express it, e.g. for the first agent, via

$$\mathrm{DPEB}_1(\mathbf{q},\, \boldsymbol{\Psi}) = \sqrt{\mathbf{q}^{\mathrm{T}} \left[\boldsymbol{\mathcal{I}}^{-1}(\boldsymbol{\Psi})\right]_{1:3,1:3} \mathbf{q}} \tag{3.54}$$

$$\leq \sqrt{\mathbb{E}\{|\mathbf{q}^{\mathrm{T}}(\hat{\mathbf{p}}_1 - \mathbf{p}_1)|^2\}}. \tag{3.55}$$

Also note that for any three orthogonal DPEBs, it also holds that the square root of their sum of squares equals the position error bound, e.g. for agent 1 (changed from [118])

$$\mathrm{PEB}_1(\boldsymbol{\Psi}) = \sqrt{\mathrm{DPEB}_1^2(\mathbf{q}_1,\, \boldsymbol{\Psi}) + \mathrm{DPEB}_1^2(\mathbf{q}_2,\, \boldsymbol{\Psi}) + \mathrm{DPEB}_1^2(\mathbf{q}_3,\, \boldsymbol{\Psi})}\,, \tag{3.56}$$

where $\{\mathbf{q}_1, \mathbf{q}_2, \mathbf{q}_3\}$ can be any orthonormal basis of $\mathbb{R}^3$.

The DPEBs in the directions of the orthogonal eigenvectors of $[\boldsymbol{\mathcal{I}}^{-1}]_{1:3,1:3}$ are given by the square roots of the corresponding eigenvalues. They are of particular importance as they yield the smallest and highest RMSE error bounds of a given deployment and represent the principal semi-axes of an ellipsoid. This ellipsoid is formed by the DPEBs in all directions and hence illustrates the localization uncertainty. Clearly, the direction which offers the highest directional information is the one that yields the smallest RMSE error bound and the same holds the other way around. Based on this geometric interpretation, we further consider the axial ratio of the DPEBs, e.g. for agent 1, as

$$\mathrm{DPEB}_1 \text{ ratio}(\boldsymbol{\Psi}) = \frac{\max_q \mathrm{DPEB}_1(\mathbf{q},\, \boldsymbol{\Psi})}{\min_q \mathrm{DPEB}_1(\mathbf{q},\, \boldsymbol{\Psi})}\,. \tag{3.57}$$

This ratio is a measure of asymmetry of the uncertainty ellipsoid, i.e. DPEB_1 ratio $= 1$ means the ellipsoid coincides with a sphere. High ratios mean that the ellipsoid's main semi-axis is significantly larger than its minor semi-axis, which implies that the overall position error bound is dominated by the estimation errors in the direction of the main semi-axis, i.e. $\mathrm{PEB}_1(\boldsymbol{\Psi}) \approx \max_q \mathrm{DPEB}_1(\mathbf{q},\, \boldsymbol{\Psi})$.

Lastly, another interesting quantity for a single agent-anchor pair m, n is the directional information in the direction of departure $\mathbf{u}_{m,n}$ ((cf. Sec. 2.2). This scalar quantity coincides with the ranging information intensity of [118] and the square-root of its inverse is what we call **D**istance **E**rror **B**ound (DEB), i.e.

$$\mathrm{DEB}_{m,n}(\boldsymbol{\Psi}) = \sqrt{\left(\mathbf{u}_{m,n}^{\mathrm{T}} \left[\boldsymbol{\mathcal{I}}(\boldsymbol{\Psi})\right]_{1:3,1:3} \mathbf{u}_{m,n}\right)^{-1}}\,. \tag{3.58}$$

The DEB is the Cramér-Rao lower bound on any corresponding unbiased distance estimator in case both coil orientations and the direction of departure are assumed known. Compared to the DPEB in the same direction, this quantity exists as long as the observation provides any spatial information in direction of $\mathbf{u}_{m,n}$, even when $1 \leq \operatorname{rank}(\boldsymbol{\mathcal{I}}(\boldsymbol{\Psi})) < 3$.

Chapter 4

Range Estimation and Localization

This chapter examines localization and ranging behavior in MI networks based on the system model from Cpt. 3. It further characterizes the underlying scaling behavior with help of the CRLB. Lastly, it highlights inherent performance bottle necks for magnetic near field systems which stem from the coupling mechanism itself and may degrade the position RMSE by orders of magnitude.

4.1 Range Estimation for Active and Passive Agents

As indicated by (3.51) and (3.52), the overall localization is ultimately a combination of the spatial and orientational information which is collected by all anchors. For MI localization, the spatial information clearly depends significantly on the range between the coils. In a first step, we thus analyze how the distance estimation quality of a single agent-anchor coil pair changes, when the agent is either active or passive. If we conceptually ignore the matching networks on both sides, it is intuitive via (2.1) that the induced current (and voltage) from an active agent to a passive anchor is defined by the mutual inductance, whereas the current induced from an active anchor to a passive agent and back is defined by the squared mutual inductance. For such a non-strongly coupled coil pair m, n at distance $d_{m,n}$ for which the dipole approximation applies, we hence expect $i_{\mathrm{T}}^{\mathrm{in}} \propto d_{m,n}^{-3}$ and $i_{\mathrm{R}}^{\mathrm{in}} \propto d_{m,n}^{-6}$, respectively. These relationships in turn imply $\mathrm{DEB} \propto d_{m,n}^{4}$ when using active agents and $\mathrm{DEB} \propto d_{m,n}^{7}$ when using passive agents (cf. (3.50) and (3.58)).

To further illustrate this matter, in Fig. 4.1a we look at an agent-anchor coil pair with variable distance $d_{m,n}$ between the coil centers and random but known orientations of the coil pair, i.e. the orientation vector of either coil is a uniform-randomly chosen point on the unit sphere. The parameters used for this simulation are given in Tab. 4.1 and the low-complexity two-port power matching strategy described in Sec. 3.3.1 is applied. Our choice for the noise parameters of the LNAs is identical to that of [28]. In Fig. 4.1b we show the resulting DEB for increasing distances $d_{m,n}$ of this setup for an either active or passive agent. The solid line represents the median DEB at each distance and the transparent area marks its interdecile range. To verify our previous explanation, we also show two lines that are proportional to $d^4_{m,n}$ and $d^7_{m,n}$, respectively.

We find that our expectations match the ranging behavior for roughly $d_{m,n} \geq 2D^{\text{coil}} = 2\,\text{cm}$. For smaller distances at which a strong coupling occurs, we observe a drastic change in the behavior of the distance estimation which results in an almost constant DEB. In this regime, active and passive agents also yield a comparable median DEB. Additionally, the figure shows how severely misalignments of a coil pair may affect the current gains and hence degrade the DEB, even for known coil orientations that do not need to be jointly estimated. For the distance estimation of the passive agents, the misalignment factor $J_{m,n}$ (or misalignment loss) is further squared due to the bidirectional use of the channel. The PDF of the alignment factor $J_{m,n}$ for a single coil pair with random orientations of both coils (uniformly random on the unit sphere) was derived in [28] and is given by

$$f_{J_{m,n}}(J_{m,n}) = \frac{\operatorname{arcosh}(2)}{\sqrt{3}} \begin{cases} 1 & |J_{m,n}| < \frac{1}{2}\,, \\ 1 - \frac{\operatorname{arcosh}(2|J_{m,n}|)}{\operatorname{arcosh}(2)} & \frac{1}{2} \leq |J_{m,n}| \leq 1\,, \\ 0 & 1 \leq |J_{m,n}| \end{cases} . \tag{4.1}$$

The PDF[1] of the alignment factor's magnitude is shown in Fig. 4.2 with highlighted percentile ranges and matches the misalignment loss of Fig. 4.1b. Overall, MI distance estimation hence demonstrates fundamentally different behaviors depending on the extent of the coils' coupling and whether active or passive agents are used.

Fig. 4.1c illustrates how the median input coupling impedance Z^{in}_{C} (note $Z^{\text{in}}_{\text{C}} = Z^{\text{out}}_{\text{C}}$ for the active case) and the median power P_{T} (active) or P_{R} (passive) into the network change for different agent-anchor distances. In the strongly-coupled regime, there is a mutual detuning of the coupling input impedances that results from an increase of the coupling input resistances (dashed lines) while the reactances (dotted lines) remain constant. Due to the imperfect matching networks, this detuning degrades the power (dashdotted lines) P_{T} (active case) or P_{R} (passive case) into the network and in turn causes a worsening of the DEB. Aside from the impact of the matching networks, the behavior of the DEB would also change when the coils get closer since the mutual inductance itself converges to a constant value for close coil proximities.

Lastly, the median DEB over the agent-anchor distance is shown for different quality factors of the involved coils in Fig. 4.1d. The quality factors are varied via the coil

[1]Part of the PDF is omitted to allow for a logarithmic scale.

agent current	i_{T}	1 mA
anchor current	i_{R}	1 mA
coil turns	N^{coil}	10
coil diameter	D^{coil}	10 mm
coil height	H^{coil}	8 mm
wire diameter	D^{wire}	0.5 mm
design frequency	f^{des}	1 MHz
op. frequency	f	1 MHz
conductivity	σ^{wire}	$59.6\,\mathrm{MS\,m^{-1}}$
rel. permittivity	ϵ_{r}	1
rel. permeability	μ_{r}	1
bandwidth	B	5 kHz
temperature	T	300 K
LNA noise variance	$(\sigma^{\mathrm{LNA}})^2$	$2 \cdot 10^{-22}\,\mathrm{A^2\,Hz^{-1}}$
LNA noise resistance	R^{LNA}	$40\,\Omega$
LNA noise corr. coeff.	ρ^{LNA}	$0.5 + 0.7j$

(a) Specified parameters.

resistance	R	$0.1\,\Omega$
self-inductance	L	$0.7\,\mu\mathrm{H}$
self-capacitance	C	0.3 pF
wire length	l^{wire}	0.3 cm
wavelength	λ	300 m
coil Q-factor at f^{des}	Q	40
coil self-resonance	f^{self}	340 MHz

(b) Resulting parameters.

Table 4.1: Simulation parameters and resulting quantities.

antenna resistances R, while the other parameters such as the geometric structure, the inductance, or the self capacitance remain constant. We find that an increasing quality factor simply shifts the same coupling behavior to higher distances. As a result, strong coupling and the impedance detuning already occur at distances $d_{m,n} \geq 2D^{\mathrm{coil}} = 2\,\mathrm{cm}$ for high quality factors.

4.2 Scaling Behavior for Practical Design Parameters

Before we examine the scaling behavior for more complex networks, let us first have a closer look at a single agent-anchor coil pair m, n. We only consider an active agent and assume that the coils are separated by a sufficiently large distance such that we can rely on the dipole approximation (2.3) of the mutual inductance and the unilateral assumption of weakly-coupled links. We additionally assume ideal power matching and neglect the coils' self capacitances such that (3.45) applies, i.e.

$$i^{\mathrm{in}} = \frac{j\omega M_{m,n}}{\sqrt{4R_m R_n}} \frac{i_{\mathrm{T}}}{2} \tag{4.2}$$

$$\approx \frac{j\mu_0\mu_{\mathrm{r}}\pi^2}{2^5\sqrt{R^{\mathrm{ref}} K_m^{\mathrm{ohm}} K_n^{\mathrm{ohm}}}} \cdot \frac{J_{m,n}}{d_{m,n}^3} \cdot \underbrace{f \cdot \sqrt{N_m^{\mathrm{coil}} N_n^{\mathrm{coil}} (D_m^{\mathrm{coil}} D_n^{\mathrm{coil}})^3 P_{\mathrm{T}}}}_{s_{m,n}} \tag{4.3}$$

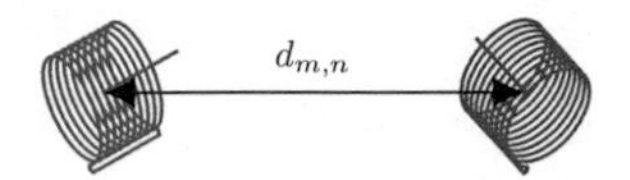

(a) Setup of agent-anchor coil pair.

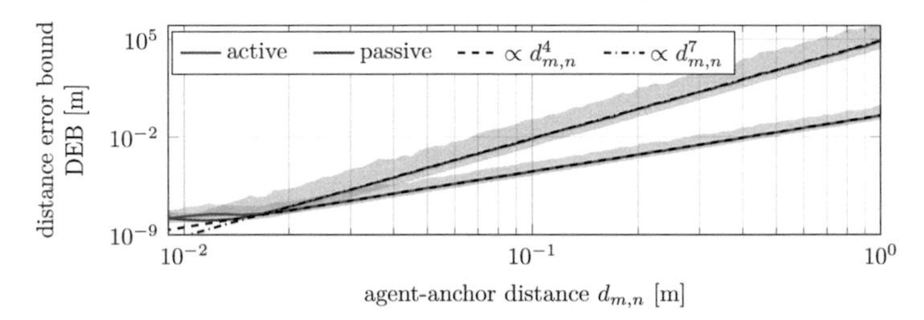

(b) DEBs for the active and passive case.

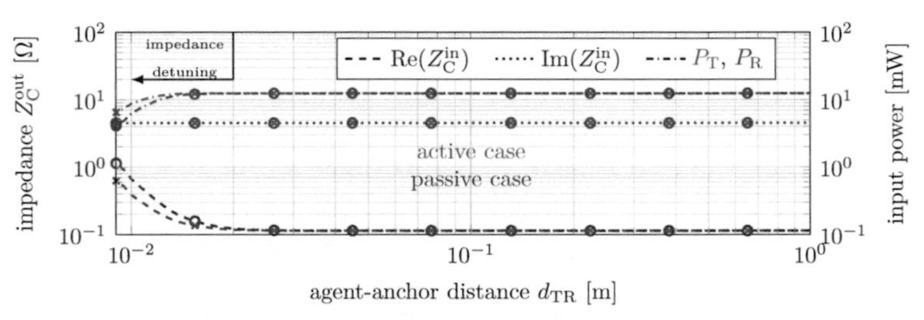

(c) Impedances and input power for the active and passive case.

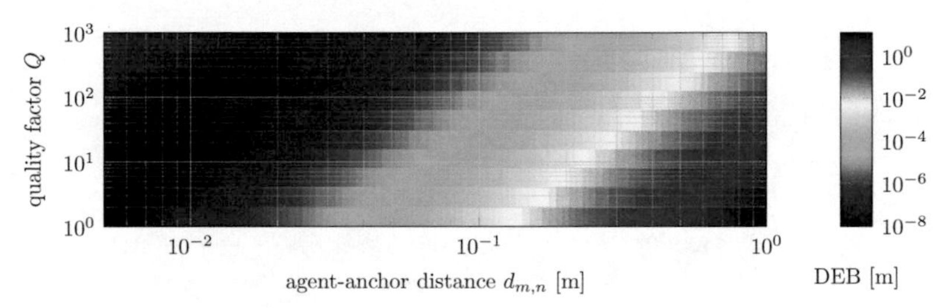

(d) Median DEB with different quality factors for the active case.

Figure 4.1: Single agent-anchor coil pair at variable distances $d_{m,n}$ with uniformly and randomly distributed orientation vectors on the unit sphere. The median DEB (solid line) and the interdecile range (transparent area) of the DEB are shown for an either active or passive agent. The corresponding input coupling impedances (left ordinate) and the power into the network (right ordinate) are also illustrated. Lastly, the impact of the quality factors on the DEB is also shown for the active case.

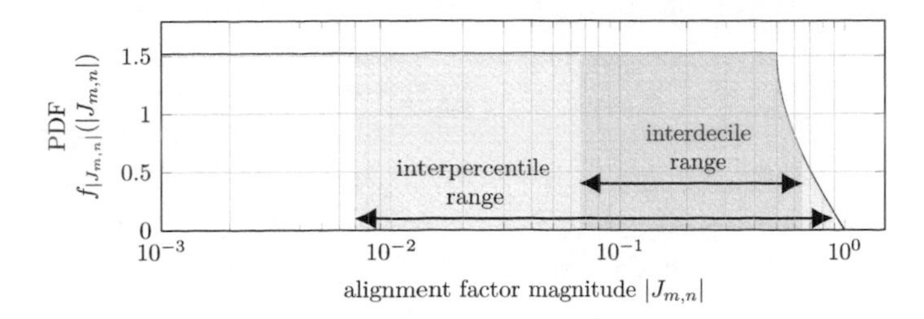

Figure 4.2: PDF of the alignment factor's magnitude for a randomly oriented coil pair.

The approximation (4.3) follows if we additionally assume that the skin effect, the proximity effect and coil height are negligible, which results in (2.9) being approximable by $R \approx D^{\text{coil}} N^{\text{coil}} K^{\text{ohm}}$ with $K^{\text{ohm}} = \frac{4}{\sigma D^{\text{wire}}}$. Moreover, we also used $P_{\text{T}} = \frac{R^{\text{ref}}|i_{\text{T}}|^2}{4}$ which follows from ideal power matching in conjunction with (3.12) and further limited ourselves to strictly real currents, i.e. $i_{\text{T}} = |i_{\text{T}}| = 2\sqrt{\frac{P_{\text{T}}}{R^{\text{ref}}}}$. Under these assumptions, we hence find that the current as well as its derivative are directly affected by the same scaling factor, that is $i^{\text{in}} \propto s_{m,n}$ and $\frac{\partial i^{\text{in}}}{\partial [\boldsymbol{\Psi}]_j} \propto s_{m,n}$. This scaling factor incorporates major practical design parameters such as the coil dimensions, the transmit power and the operating frequency[2]. It e.g. shows that a halving of the transmit power may be counteracted by increasing the anchor coil diameter by $\sqrt[3]{2}$. Via (3.50) this also means that the FIM of every independent agent-anchor link behaves according to $\boldsymbol{\mathcal{I}} \propto s_{m,n}^2$. If a network constellation is fixed and all identical anchor coils provide independent observations, we hence find that the PEB is characterized by PEB $\propto s_{m,n}^{-1}$ (cf. (3.52) and (3.51)).

In order to get an intuition on the potential and the limitations of the approximation PEB $\propto s_{m,n}^{-1}$ we study the coplanar sensor network depicted in Fig. 4.3a comprising $N_{\text{T}} = 1$ active agent and $N_{\text{R}} = 4$ passive anchors. For this setup, we compare the exact PEB with the one obtained with the approximation if different parameters are varied individually. The default system parameters are again specified by Tab. 4.1 and ideal power matching networks are used, which leads to the default transmit power of $P_{\text{T}} = 12.5\,\mu\text{W}$. Since all coils are identical, the impact of the coil diameter and the number of coil windings is squared. Fig. 4.3b and Fig. 4.3c show the corresponding PEBs if we scale the design parameters individually. That is, X_0 denotes the above choice for the corresponding parameter and X denotes the scaled version while all other parameters remain unchanged. The simulations ignore all of the above assump-

[2]The design frequency f^{des} is scaled in correspondence with the operating frequency f.

tions (*full model*), so they e.g. consider imperfect power matching networks and back coupling of the anchors. The results are further compared to the expected scaling behavior according to the derived scaling factor ($\propto s_{m,n}^{-1}$). Despite the numerous simplifying assumptions that were necessary to obtain the scaling factor, we see that it characterizes the impact of the transmit power and the coil diameter well. However, when scaling the operating frequency a diverging behavior can be observed due to the skin effect: For lower frequencies, the increase of the PEB does not keep up with our expectations, since this effect is mitigated by the unconsidered decrease of both coil resistances. For higher frequencies the contrary can be observed and the PEB roughly follows PEB $\propto \frac{1}{\sqrt{f}}$ instead of PEB $\propto \frac{1}{f}$. We observe a similar discrepancy when changing the number of coil windings, which is a result of the proximity effect: For fewer coil windings the proximity effect is less distinct so the coil resistances decrease more significantly than expected. When increasing the number of coil windings the PEB suffers from the higher resistances due to the amplified proximity effect. Yet, since the proximity effect itself converges for a higher number of coil windings [27], the discrepancy between the expected PEB and the actually observed one is negligible. To emphasize these statements, we also show the PEB which is obtained when omitting either the skin effect (*no skin effect*) or the proximity effect (*no proximity effect*) from the simulation. For these special cases, the lines are offset by a fixed constant such that they intersect at $X/X_0 = 1$. Overall, we conclude that the approximated scaling factor is well-suited to asses the impact of different transmit powers, coil diameters or, to some extent, even the numbers of coil turns. However, for the operating frequency it shows to be a bad fit as the skin effect has a significant impact on the coil resistance, even on the considered frequencies in the interval $f \in [0.110]$MHz. Moreover, since the scaling factor was derived for the dipole approximation and a weak coupling, it is unlikely to be accurate for dense networks with close coil proximities.

4.3 Impact of Anchor Placement and Density

Another common design parameter of interest for the active agent case, which is not directly evident from (4.3), is the total number N_R of anchors used. However, even with all assumptions from Sec. 4.2 its impact is hard to characterize since it is highly dependent on the network topology itself. Still, the unilateral assumption and the ideal power matching lead to $\mathbf{K} \propto \mathbf{I}_{N_\mathrm{R}}$ and hence $\frac{\partial \mathbf{K}}{\partial [\mathbf{\Psi}]_j} = \mathbf{0}_{N_\mathrm{R}}$, so we can consequently write

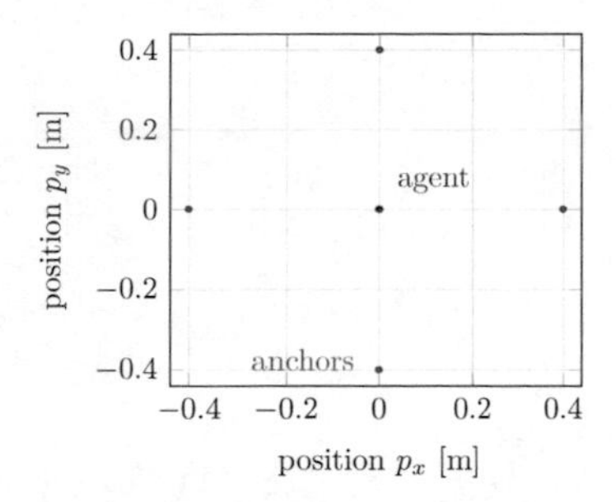

(a) Coplanar coil setup with equidistant and well-spread out anchor coils.

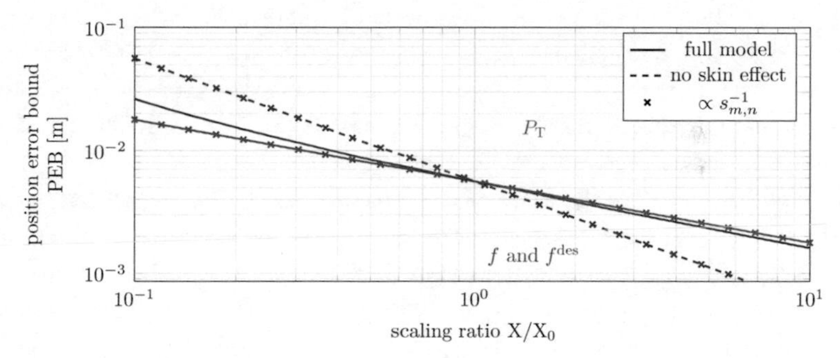

(b) Scaling behavior for changes in the transmit power and used frequencies.

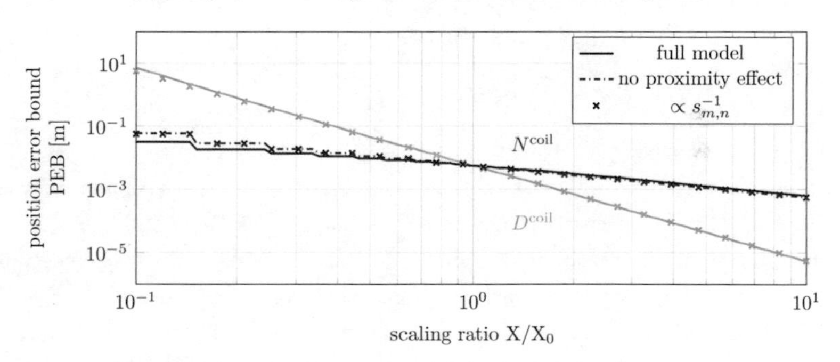

(c) Scaling behavior for changes in the coil diameter and number of coil windings.

Figure 4.3: Scaling behavior analysis for common design parameters in a fixed setup with $N_\mathrm{T} = 1$ agent and $N_\mathrm{R} = 4$ equidistant and well-spread out anchors. The figure compares the performance when using the full model to the expected behavior according the simplified scaling factor. Two special cases which neglect the skin effect and the proximity effect are also shown.

the FIM as (cf. (3.50) and (3.51))

$$\boldsymbol{\mathcal{I}} \propto \mathrm{Re}\left(\begin{bmatrix} \left\|\frac{\partial \mathbf{i}^{\mathrm{in}}}{\partial \psi_1}\right\|^2 & \frac{\partial (\mathbf{i}^{\mathrm{in}})^{\mathrm{H}}}{\partial \psi_1}\frac{\partial \mathbf{i}^{\mathrm{in}}}{\partial \psi_2} & \cdots \\ \frac{\partial (\mathbf{i}^{\mathrm{in}})^{\mathrm{H}}}{\partial \psi_2}\frac{\partial \mathbf{i}^{\mathrm{in}}}{\partial \psi_1} & \left\|\frac{\partial \mathbf{i}^{\mathrm{in}}}{\partial \psi_2}\right\|^2 & \cdots \\ \vdots & \vdots & \ddots \end{bmatrix}\right) \tag{4.4}$$

$$= \sum_{n_\mathrm{R}=1}^{N_\mathrm{R}} \mathrm{Re}\left(\begin{bmatrix} \left|\frac{\partial i^{\mathrm{in}}_{n_\mathrm{R}}}{\partial \psi_1}\right|^2 & \frac{\partial (i^{\mathrm{in}}_{n_\mathrm{R}})^*}{\partial \psi_1}\frac{\partial i^{\mathrm{in}}_{n_\mathrm{R}}}{\partial \psi_2} & \cdots \\ \frac{\partial (i^{\mathrm{in}}_{n_\mathrm{R}})^*}{\partial \psi_2}\frac{\partial i^{\mathrm{in}}_{n_\mathrm{R}}}{\partial \psi_1} & \left|\frac{\partial i^{\mathrm{in}}_{n_\mathrm{R}}}{\partial \psi_2}\right|^2 & \cdots \\ \vdots & \vdots & \ddots \end{bmatrix}\right). \tag{4.5}$$

which directly highlights the additivity property for Gaussian independent observations. Interpreting (4.5), we see that every anchor provides additional spatial information depending on how its input current varies for changes of the agent deployment. For the coplanar setup of Fig. 4.3a, agent displacements along the radial direction of an agent-anchor pair (i.e. in each respective direction $\mathbf{u}_{m,n}$) directly affect the the distance $d_{m,n}$ and hence have a significant impact on the input current. In comparison, agent displacements along the perpendicular direction $(\mathbf{u}_{m,n})_\perp$ have a negligible impact on the distance and hence the input currents. For each individual agent-anchor pair there is thus a lack of spatial information in the direction $(\mathbf{u}_{m,n})_\perp$. Lastly, under the very special assumption that each anchor provides the same spatial information, i.e. the anchors have the same placement or are located in specific symmetric constellations, we find $\boldsymbol{\mathcal{I}} \propto N_\mathrm{R}$ and thus PEB $\propto N_\mathrm{R}^{-\frac{1}{2}}$ with respect to the number of anchors. While this assumption generally does not hold, it shows that even without novel spatial information, the anchors may contribute to a decreasing PEB via noise-averaging effect.

In Fig. 4.4, we look at a coplanar network topology and analyze the obtained PEB for an increasing number of anchors. That is, we now place the anchors uniformly random either on the outer circle shown in Fig. 4.3a, the middle circle, or within the annulus formed by those two circle, while enforcing a minimum distance constraint of $d^{\mathrm{min}} \geq 2D^{\mathrm{coil}}$ for any pair of anchors. Again, the solid lines present the median obtained PEB and the transparent areas mark the corresponding interdecile ranges. Additionally, we show two dashed lines which are both proportional to $N_\mathrm{R}^{-\frac{1}{2}}$. For both circle placements, we see that there is a strong variability of the PEB for $N_\mathrm{R} < 5$. For these operating points, the exact placement of the anchors, or rather the directions which are covered by each anchor, play a crucial role for the PEB. Some additional anchors hence provide localization benefits far beyond that of the noise-averaging effect.

Starting at $N_\mathrm{R} = 5$ anchors, this variability is already reduced significantly and the exact placement of new anchors is almost irrelevant for the PEB. Further increasing the number of anchors shows that any random constellation only reduces the PEB according to $N_\mathrm{R}^{-\frac{1}{2}}$. For the annulus placement a similar behavior is observed, however more anchors are required to meaningfully decrease the interdecile range of the PEB. This phenomenon occurs as the anchors now do not only need to cover the different directions well, but also need to be placed as close to the agent as possible. Statistically, it hence takes more randomly placed anchors to saturate these favorable placement areas. Moreover, for higher anchor numbers the probability of placing all of them close to either circle is very low, which is why the interdecile range of the annulus placement does not overlap with the interdecile ranges of the circular anchor placements.

In comparison, Fig. 4.4c shows a scenario where the anchors are placed on the inner circle when reducing the minimum distance constraint between anchors to $d^\mathrm{min} \geq D^\mathrm{coil}$, which means that at most $N_\mathrm{R} = 40$ coils can be placed. For this scenario, we still observe comparable results for low agents numbers, which then however switch to a contrasting behavior. Namely, the PEB starts to degrade for a further increasing number of anchors due to their mutual impedance detuning. This is further illustrated by the dotted line, which shows the median PEB in case all identical anchors are decoupled. For this decoupled anchor case the median PEB is comparable to the ones in Fig. 4.4 and no deteriorating effect is apparent. Additionally, we also show the PEB that is obtained when the anchors on the inner circle are well-spread, i.e. they are placed on the N_R vertices of the corresponding regular inscribed polygon while still satisfying the minimum distance constraint $d^\mathrm{min} \geq D^\mathrm{coil}$. This well-spread constellation constitutes the optimal anchor placement in case the observations only depend on the distances of the individual agent-anchor pairs [119]. Yet, as we consider coupling between the anchors, mutual impedance detuning, and imperfect matching networks, there are random constellations that slightly outperform this well-spread constellation.

Overall, the optimal anchor placement requires optimization of the spatial diversity, i.e. the directional information as well as the optimization of the range measurement quality [119, 120]. We see that this also holds for MI localization, and it is hence advantageous to (i) have the anchors placed closely to the area which the agent occupies to increase the range measurement quality and (ii) to be equally spread in order to balance the directional information in all dimensions. In [121] this was also demonstrated for time-of-arrival based localization , when the agent locations have a truncated radially-symmetric PDF and when the anchors may be placed within an surrounding annulus. In our scenarios, further increasing the number of anchors quickly shows no further benefits other than the noise-averaging effect. When the anchors have a close proximity, this noise-averaging effect also shows diminishing returns and can even be detrimental due to the mutual detuning of the anchor impedances which is associated with a worsening of the receive SNR.

4.4 Impact of Agent Placement

The previous investigation is interesting for initial considerations about the optimal anchor placement. Its findings about directional information and range measurement quality still apply, when the anchor placements are fixed and the agent can move freely. However, this opposite perspective better highlights the limitations of conventional MI localization and the limited application area. In Fig. 4.5 we hence show intensity plots of the PEB and the DPEB ratio for various coplanar agent placements while the anchors are fixed according to Fig. 4.3a. In detail, Fig. 4.5a shows a close-up of the PEB for different agent placements that are in proximity to the anchors, whereas Fig. 4.5c also shows the PEB for more distant placements of the agent. Fig. 4.5b and Fig. 4.5d are the analogous plots for the corresponding DPEB ratio. We first distinguish two different scenarios:

Clustered Anchors In case the agent m is placed far outside the anchor area, i.e. $\|\mathbf{p}_m\| \gg 1.5\,\mathrm{m}$, all direction vectors $\mathbf{u}_{m,n}$ from the anchors to the agent are approximately equal. The full anchor array hence starts to appear clustered and acts similar to a single anchor placed at the origin. This is also indicated by the roughly circularly-symmetric behavior of the PEB, which is apparent in Fig. 4.5c. While coplanar localization is still possible, the DPEB in the tangential direction of this circle is significantly higher than the one in the radial direction, regardless where on the two-dimensional plane the agent is placed. This asymmetry of the DPEBs is also indicated by Fig. 4.5d, which shows DPEB ratio ≥ 10 for the clustered anchor scenario. Thus, the overall PEB is dominated by uncertainties in this tangential direction. For three-dimensional network constellations and random coil orientations, this effect is usually still observable. However, it is sometimes slightly mitigated in case tangential movements cause a more beneficial alignment of the coils.

Distributed Anchors Conversely, if the agent is still placed in the vicinity of the anchor array with $\|\mathbf{p}_m\| \leq 1.5\,\mathrm{m}$, the geometric structure of the anchors relative to the agent plays a crucial role and defines the obtainable localization accuracy. If

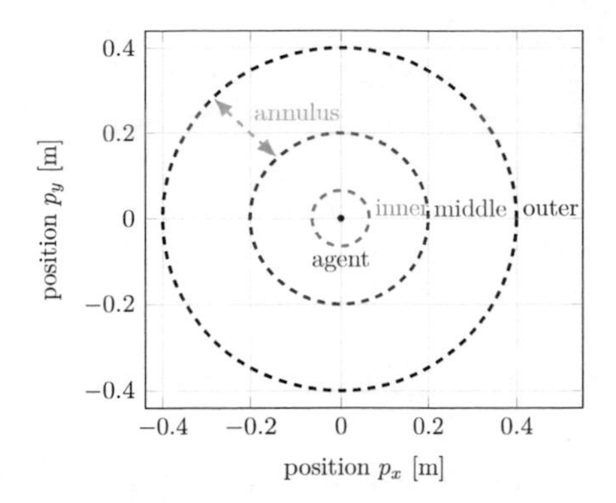

(a) Coplanar coil setup with N_R anchor coils on the surrounding structures.

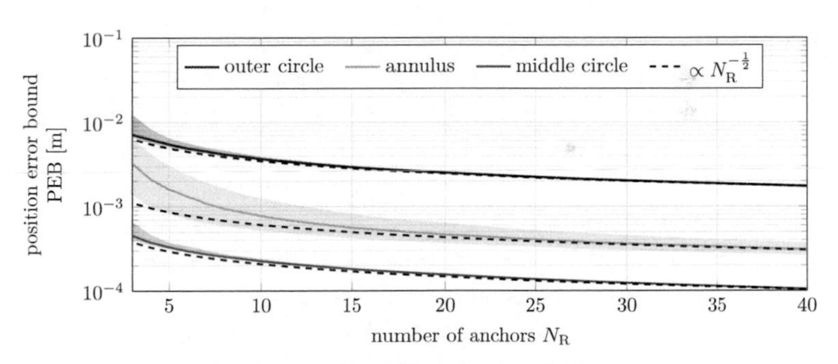

(b) PEB for outer circle, middle circle and annulus placement.

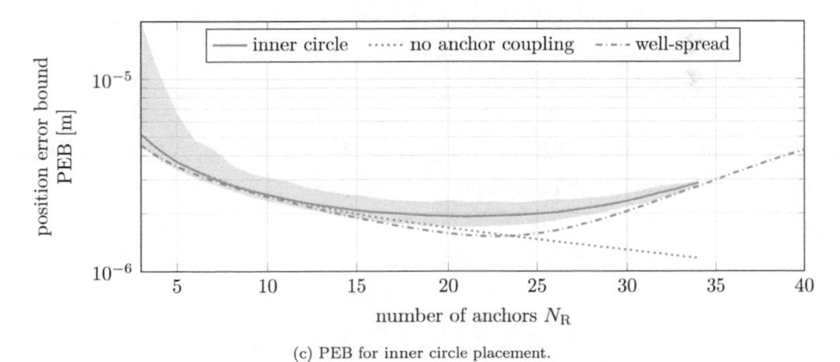

(c) PEB for inner circle placement.

Figure 4.4: Scaling behavior analysis when increasing the number of anchors N_R that are uniform randomly placed on different structures around the agent.

the agent is located at point A, all anchors are well-spread and close to the agent, so the spatial information is balanced and yields the minimum PEB as well as the minimum DPEB ratio. At point B, the agent's distance towards two of the anchors is drastically increased (compared to point A), which leads to a slightly higher PEB and DPEB ratio. Yet, radial and tangential movements still cause significant changes of the observed input currents at the two close anchors, so the agent can be located accurately. In comparison, at point C tangential movements of the agent have almost no impact on the input currents of the anchors that are placed at the abscissa. The spatial information in the tangential direction thus has to be provided by the anchors located on the ordinate, which however have a higher distance and thus a worse SNR. The overall PEB at point C is consequently already dominated by the uncertainty in the tangential direction, while its placement in the radial direction can be estimated well.

Both of these scenarios are of interest, as the placement of the anchors may in reality not be fully flexible but rather constraint by the practical limitations of the desired application, e.g. due to a required minimum sensor size or other hardware requirements. Additionally, the PDF of the agent deployment may not be known beforehand, so an optimized anchor placement may be unrealistic. This could occur for medical in-body application, where the agent may be traveling through the gastrointestinal tract and the anchors have to be placed in a clustered sensor head that is used outside of the body. The optimization of the anchor placement would hence be highly limited. Moreover, the area of interest with respect to the agent deployment might depend on the medical cause that is being screened for, so it would be unknown beforehand.

4.5 Conclusions

For the employed system model, we showed that MI localization exhibits a drastically different behavior when the coils are in close proximity and strongly coupled, compared to when they are further apart and weakly coupled. For the weakly-coupled regime, we demonstrated that the distance estimation of passive agents becomes unfeasible quickly, as the underlying distance estimation degrades according to DEB $\propto d_{m,n}^7$ with $d_{m,n}$ as distance between an passive agent and active anchor. In contrast, the distance estimation of active agents works well in this regime as they are less affected by the increasing distances of the coils, i.e. DEB $\propto d_{m,n}^4$. For the strongly-coupled regime, we also showed that the coils cause a mutual impedance detuning, which impairs the

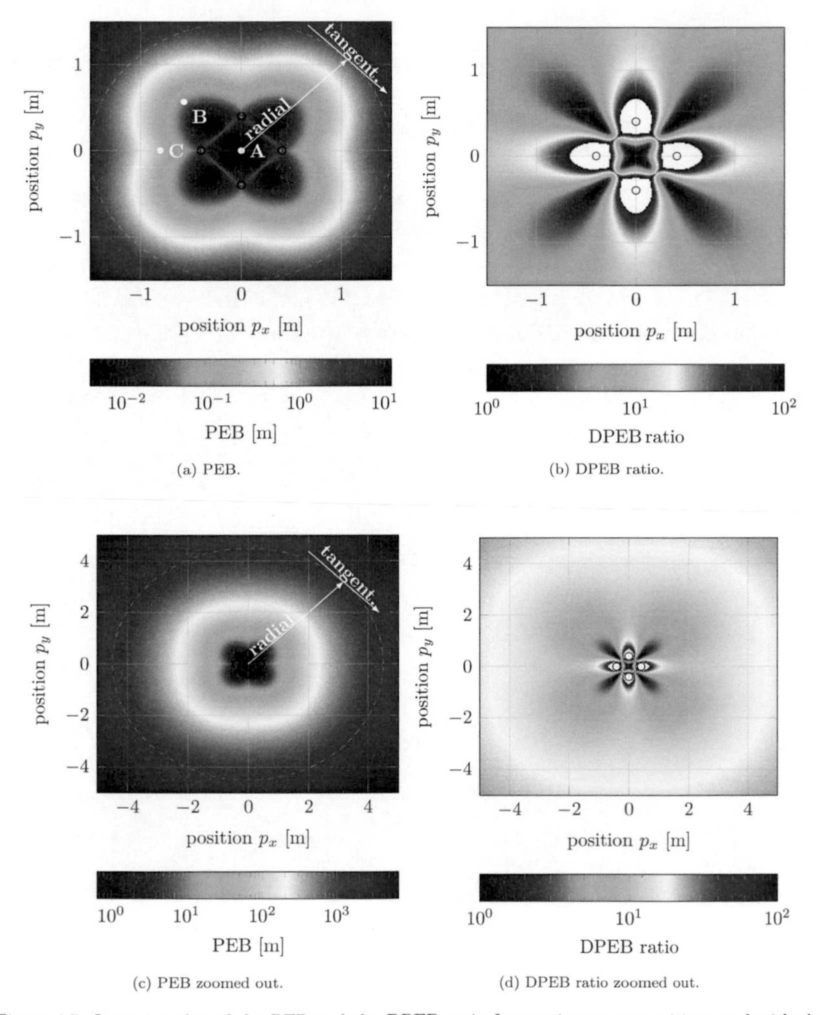

(a) PEB.

(b) DPEB ratio.

(c) PEB zoomed out.

(d) DPEB ratio zoomed out.

Figure 4.5: Intensity plot of the PEB and the DPEB ratio for varying agent positions and with the fixed anchor placement of Fig. 4.3a. The anchors are represented by black circles and are not true to scale. For visibility purposes, we have also cut off the color scale for DPEB ratios larger than 100.

transmit power and power transfer, leading to a deterioration of the distance estimation. For active agents in the weakly-coupled regime, we further characterized the impact of practical system parameters such as the diameter of the involved coils and offered a scaling factor as simple means to adjust the system design to ones individual application requirements.

Additional investigations of selected network constellations revealed that MI localization can be highly accurate if the following criteria are met simultaneously: (i) the agent-anchor coupling has to be strong enough to not suffer from a drastic path loss, but not so strong that the coils trigger a mutual impedance detuning, (ii) the anchors need to provide orientational diversity to mitigate misalignment losses, and (iii) the anchors need to be spatially distributed such that the obtained directional information about the agent position is balanced. If practical limitations make these requirements unattainable, MI localization systems quickly become inaccurate and unreliable.

Chapter 5

Passive Relays and Relay-Aided Localization

Note: Parts of this chapter have been published by us in [93]. This work hence exhibits similarities regarding formulations and visualizations.

In this chapter, we propose the use of a new class of sensor nodes called *passive relays* in order to enhance MI localization and to mitigate the issues described in Sec. 4.5, namely the range limitation, the misalignment loss and the asymmetry of directional information. These sensors are supposed to function as auxiliary nodes that are ubiquitous within the network. To this end, they are always passive, i.e. they do not require any sources. Moreover, their deployment is assumed to be known. Due to their low hardware complexity, passive tags are cheap to produce and can easily be distributed within a given area, e.g. as smart dust [31]. Passive relays have already been studied for communication and WPT, e.g. in [19, 32, 122–125]. However, so far their main use has been to increase channel gains to boost communication capabilities either by providing orientational diversity or by establishing a waveguide effect. Yet, such a waveguide effect requires specific alignments of all passive relays and simply using arbitrary alignments can lead to a degradation of the channel [32]. Apart from their beneficial impact on the channel gains, passive relays have also been studied for a single agent-anchor coil pair in [31]. In this work it was shown in that using passive relays which switch their loads can lead to independent measurements at the anchor. This process in turn resolved position ambiguities of the agent and thus enabled single-anchor localization. Except for these works, passive relays have not been studied thoroughly for localization to the best of our knowledge.

The goal of this chapter is to show that passive relays can have further advantages for localization, other than only providing stronger channel gains or enabling single-anchor localization. More precisely, we want to show that they are well-suited to balance the directional spatial information of the agents. In Sec. 5.1, we summarize how to easily incorporate those type of sensors into our existing system model. In Sec. 5.2 we analyze the impact of single passive relays on the coupling mechanisms by simulation and explain how this in turn affects the localization. This study is then

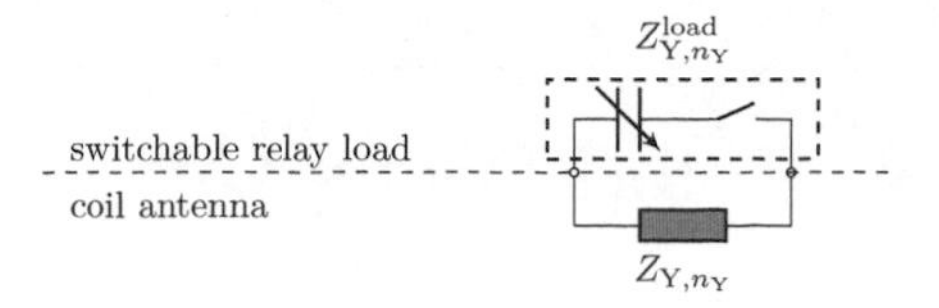

Figure 5.1: Circuit representation for a switchable passive relay coil.

extended to multiple passive relays for selected topologies to visualize all advantageous localization effects of passive relays. In Sec. 5.3 we enhance the functionality of passive relays and enable them to switch between multiple different states of their complex loads. Based on this enhancement we propose multiple practical switching schemes. We then compare how these switching schemes improve the localization accuracy and range in arbitrary MI networks.

5.1 Incorporation of Passive Relay Coils

The equivalent circuit of a single passive relay n_Y is shown in Fig. 5.1. It consists of a coil antenna represented by its complex impedance Z_{Y,n_Y}, a variable load and a switch. The series impedance (of the switch and the actual load) takes on a finite value with $\text{Re}\{Z^{\text{load}}_{\text{Y},n_\text{Y}}\} \geq 0$ for a closed switch and $Z^{\text{load}}_{\text{Y},n_\text{Y}} = Z_\infty$ for an open switch.

We assume that all deployments of these new nodes are known and that we can intentionally control whether to turn them *on* (closed switch) or *off* (open switch). Note that the switch state *off* implies that no current flows in the respective relays, so they do not impact the remaining network as long as their self capacitances can be neglected. In contrast, all relays in the *on* state carry induced currents and therefore generate additional fields, which alter the input currents seen at the anchors. Consequently, the mere presence of the passive relays will affect the coupling impedance matrix $\mathbf{Z}_\text{C}$ and thus our anchor observations. As before, we can characterize the near-field coupling between all agents, anchors and passive relays by a multiport (cf. Fig. 2.4), i.e. $N = N_\text{T} + N_\text{R} + N_\text{Y}$ with N_Y being the number of passive relays. Since the current-voltage relationships at all passive ports are predetermined by their loads, all passive relay ports can be collapsed to obtain a new coupling matrix $\tilde{\mathbf{Z}}_\text{C}$ between agents and anchors, which additionally incorporates the relays' impact. More precisely, in case there are passive relays in the network, we have to extend the coil antenna impedance matrix to [28]

$$\begin{bmatrix} \multicolumn{2}{c}{\multirow{2}{*}{\mathbf{Z}_\mathrm{C}}} & \mathbf{Z}_\mathrm{C:YT}^\mathrm{T} \\ & & \mathbf{Z}_\mathrm{C:YR}^\mathrm{T} \\ \mathbf{Z}_\mathrm{C:YT} & \mathbf{Z}_\mathrm{C:YR} & \mathbf{Z}_\mathrm{C:Y} \end{bmatrix} \in \mathbb{C}^{N\times N}. \tag{5.1}$$

with analogous construction to $\mathbf{Z}_\mathrm{C}$. The impedance matrix $\mathbf{Z}_\mathrm{C:Y}$ describes the relay coupling while incorporating the inductive interactions caused by the self capacitances of all coils. The transimpedance matrices $\mathbf{Z}_\mathrm{C:YT}$ and $\mathbf{Z}_\mathrm{C:YR}$ on the other hand characterize the current-voltage relationships from the agents and anchors to the relays, respectively. As described, the relationships of currents and voltages on the relays are predetermined via their loads, the full impedance matrix of (5.1) can be collapsed to

$$\tilde{\mathbf{Z}}_\mathrm{C} = \underbrace{\mathbf{Z}_\mathrm{C}}_{\text{direct path}} \underbrace{- \begin{bmatrix} \mathbf{Z}_\mathrm{C:YT}^\mathrm{T} \\ \mathbf{Z}_\mathrm{C:YR}^\mathrm{T} \end{bmatrix} \left(\mathbf{Z}_\mathrm{C:Y} + \mathbf{Z}_\mathrm{Y}^\mathrm{load}\right)^{-1} \begin{bmatrix} \mathbf{Z}_\mathrm{C:YT} & \mathbf{Z}_\mathrm{C:YR} \end{bmatrix}}_{\text{relay path}}, \tag{5.2}$$

with $\mathbf{Z}_\mathrm{Y}^\mathrm{load} = \mathrm{diag}\left(Z_{\mathrm{Y},1}^\mathrm{load}, \ldots, Z_{\mathrm{Y},N_\mathrm{Y}}^\mathrm{load}\right)$ being the load matrix. So $\tilde{\mathbf{Z}}_\mathrm{C}$ has the same dimensions as $\mathbf{Z}_\mathrm{C}$ and also describes the current-voltage relationships between agents and anchors but now incorporates the full impact of all passive relays. In case passive relays are present in the network, $\mathbf{Z}_\mathrm{C}$ can hence simply be replaced by $\tilde{\mathbf{Z}}_\mathrm{C}$ without otherwise changing any of the previous formulations. The additionally required derivatives for the estimation bounds are stated in Appendix A.

5.2 Localization Impact of Passive Relays for Selected Topologies

In a first step, we analyze selected topologies comprising agents, anchors and passive relays to develop intuition and to better understand the impact that resonant passive relays have on MI localization and communication. For all simulations of this chapter we use the system parameters stated in Tab. 4.1 and identical solenoid coils for all nodes[1]. Further, for Sec. 5.2 the relays are always resonantly loaded. That is, the relay switch is closed and the load itself is a capacitance with impedance $Z_{\mathrm{Y},1}^\mathrm{load} = -\mathrm{Im}(Z_{\mathrm{Y},1})$, which makes the relay resonant at our operating frequency. In contrast, we consider random topologies and possibly non-resonant relays in Sec. 5.3.

[1] If three-axis coils are considered, its subcoils are identical to these solenoid coils.

5.2.1 Localization Impact of a Single Resonant Passive Relay

In a first step, we consider a coplanar clustered anchor scenario with $N_{\mathrm{R}} = 5$ anchors as shown in Fig. 5.2. A single agent is fixed at a high distance from the origin and a single relay is placed at a variable distance d_{TY} from the agent, as indicated. Fig. 5.2e shows that for distances larger than the one from the agent to point A (cf. Fig. 5.2a), the relay has no relevant impact on the localization as the corresponding PEB remains constant. In this regime and in Fig. 5.2a, we observe the same localization behavior as in Sec. 4.4 for clustered anchors, i.e. the radial direction (in this case p_x) can be estimated well while the tangential direction (in this case p_y) is hard to estimate and dominates the overall PEB. When the relay moves closer to the agent, the agent-induced relay current increases and the relay in turn starts to contribute to the observed input currents at all anchors. This effect is more intuitive when looking at the expression (5.2) for a single agent-anchor pair in case only a single relay is present and all self capacitances can be ignored:

$$\tilde{\mathbf{Z}}_{\mathrm{C}} = \underbrace{\begin{bmatrix} R_{\mathrm{T}} + j\omega L_{\mathrm{T}} & j\omega M_{\mathrm{TR}} \\ j\omega M_{\mathrm{RT}} & R_{\mathrm{R}} + j\omega L_{\mathrm{R}} \end{bmatrix}}_{\text{direct path}} + \underbrace{\frac{\omega^2}{R_{\mathrm{Y}}} \begin{bmatrix} M_{\mathrm{YT}}^2 & M_{\mathrm{YT}}^{\mathrm{T}} M_{\mathrm{YR}} \\ M_{\mathrm{YR}}^{\mathrm{T}} M_{\mathrm{YT}} & M_{\mathrm{YR}}^2 \end{bmatrix}}_{\text{relay path}}, \tag{5.3}$$

where we omitted the agent, anchor and relay numbers from the indices. The analogous relay-incorporated transimpedance between agent and anchor is thus given by $\tilde{\mathbf{Z}}_{\mathrm{C:RT}} = j\omega M_{\mathrm{RT}} + \frac{\omega^2}{R_{\mathrm{Y}}} M_{\mathrm{YR}}^{\mathrm{T}} M_{\mathrm{YT}}$ (cf. (3.2)). If we further rely on the unilateral assumption between agent and anchor in conjunction with ideal power matching, we obtain (cf. (3.45))

$$g^{\mathrm{active}} = h_{\mathrm{T}} = \underbrace{\frac{j\omega M_{\mathrm{RT}}}{\sqrt{4(R_{\mathrm{T}} + \frac{\omega^2 M_{\mathrm{YT}}^2}{R_{\mathrm{Y}}})(R_{\mathrm{R}} + \frac{\omega^2 M_{\mathrm{YR}}^2}{R_{\mathrm{Y}}})}}}_{h^{\mathrm{direct}}} + \underbrace{\frac{\frac{\omega^2}{R_{\mathrm{Y}}} M_{\mathrm{YR}}^{\mathrm{T}} M_{\mathrm{YT}}}{\sqrt{4(R_{\mathrm{T}} + \frac{\omega^2 M_{\mathrm{YT}}^2}{R_{\mathrm{Y}}})(R_{\mathrm{R}} + \frac{\omega^2 M_{\mathrm{YR}}^2}{R_{\mathrm{Y}}})}}}_{h^{\mathrm{relay}}}. \tag{5.4}$$

Since $i_{\mathrm{T}}^{\mathrm{in}} = \frac{1}{2}\, g^{\mathrm{active}}\, i_{\mathrm{T}}$ (cf. (3.44)), it is clear that the relay contribution to the input currents at each uncoupled anchor is 90° phase-shifted compared to the direct contribution of the agent. Moreover, the derivative of this contribution with respect to the agent deployment depends heavily on the mutual inductance M_{YT} between agent and relay. As a result, additional spatial information is provided by the relay. Furthermore, this spatial information is dominated by directional information in the p_y

direction and thus manages to mitigate the deficiencies of the clustered anchor topology. This effect improves the PEB and hence the RMSE, even without meaningfully affecting the channel gains. This is further emphasized by Fig. 5.3, where we show the PTE η_T from the agent to the centrally placed anchor[2]. For this plot, we enforce the unilateral assumption and ideal power matching, such that (5.4) holds. We further partition this PTE into the contribution from the direct path $\left|h^{\text{direct}}\right|^2$ and the relay path $\left|h^{\text{relay}}\right|^2$. Additionally, Fig. 5.3 also shows the transmit power of the corresponding agent $|x_T|^2$ if no assumptions are being enforced. The closer the relay gets to the agent, the stronger its contribution. As a result, the DPEB is almost perfectly balanced for $d_{TR} = 3\,\text{cm}$. At point B (cf. Fig. 5.2b), the PEB is minimized and the DPEB ratio already started to increase again, as the uncertainty in p_y direction is now even smaller than the one in p_x direction. However, at this point, the relay is close enough to start impacting the direct path, which is evident from the denominator of (5.4). As a result, the PTE contribution of the direct path starts to decrease while the contribution of the relay path continues to advance due to the ongoing increase of the mutual inductance M_{YT}. This impact can be observed for agent-relay distances between points B and C (cf. Fig. 5.2c), for which the worsened direct contribution leads to a higher DPEB in the p_x direction. For further decreasing agent-relay distances there is a noticeable detuning of the agent impedances due to the presence of the relay. For these agent-relay distances the transmit power $|x_T|^2$ decreases significantly and the ideal agent power matching assumption that was necessary to derive (5.4) cannot be justified anymore. This detuning via passive relays which are strongly coupled with the agent is what we call *proximity problem*[3] and it can nullify the beneficial impact of passive relays or even be detrimental to the overall localization. Lastly, if more than one passive relay is present, the relay contribution at each anchor is not necessarily 90° phase-shifted, but may instead be represented by a complex phasor, as shown in [32]. The magnitude and phase of this complex phasor depend on the specific network constellation and it can reduce the PTE to all anchors, additionally to the degradation caused by the impedance detuning [28]. Yet, while the PTE reduction via such a complex phasor is detrimental for communications, it is intuitive that it may still provide additional directional information of the agent and hence be beneficial for localization.

[2]Since the PTE for the SISO case is only based on the channel gain h_T (cf. (3.38)), it is affected by the passive relay's contributions to the relay-incorporated impedance matrix $\tilde{\mathbf{Z}}_C$ However, compared to the overall receive power $|y_T|^2$ it does not change for a reduced transmit power $|x_T|^2$ which may occur as a result of an impedance detuning on the agent side.

[3]Note that the proximity problem does not occur if ideal and adaptive power matching networks are being used or if the current source is able to provide a constant current into the network.

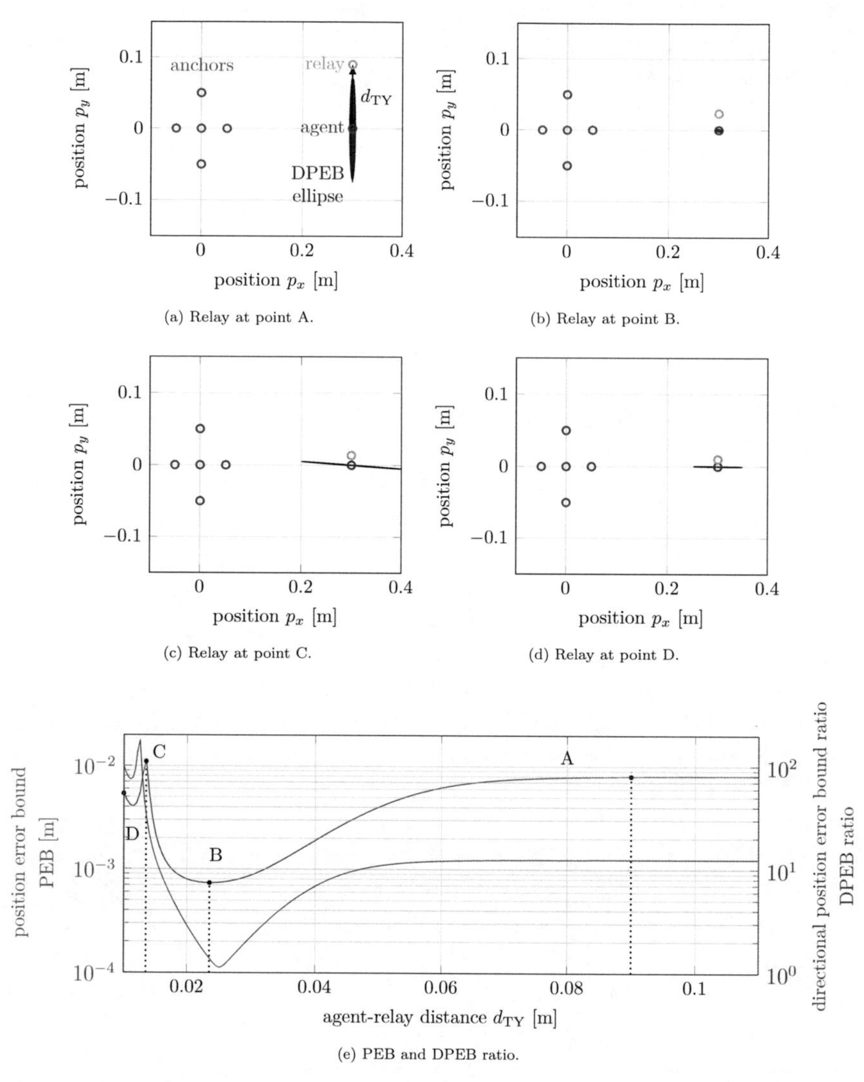

Figure 5.2: Localization impact of a resonant passive relay on a MI agent-anchor system for variable agent-relay distances. The continuous evolution of the PEB and DPEB ratio are shown in Fig. 5.2e and Fig. 5.2a to Fig. 5.2d show the corresponding network constellations and error ellipsoids for a few selected agent-relay distances. The coils and DPEB ellipses are not true to scale for better visibility.

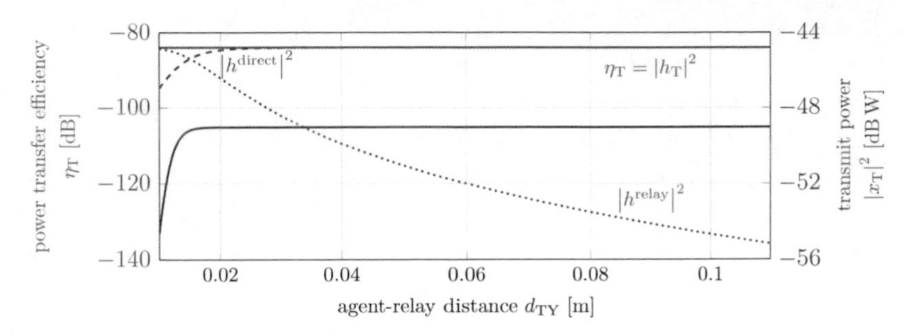

Figure 5.3: Impact of a resonant passive relay on the PTE and transmit power for the link between the agent and the centrally placed anchor of the MI system shown in Fig. 5.2.

Next, we study the impact that a passive relay has on a misaligned agent-anchor pair. To this end we consider the setup shown in Fig. 5.4a, where an agent and anchor are separated by a distance of $d_{TR} = 30\,\text{cm}$ and are almost fully misaligned with $|J_{RT}| = 0.008$. Additionally, we place a passive relay on the line connecting agent and anchor with a distance of $d_{TY} = 3\,\text{cm}$ to the agent. In Fig. 5.4b and Fig. 5.4c we further show how the DEB and PTE of this setup change for increasing rotation angles β of the relay. We also show two reference cases *NoRel* and *NoRelCopl*, for which there is no relay present. For the *NoRel* case the agent and anchor are aligned exactly as shown in Fig. 5.4a, while for the *NoRelCopl* case the agent and anchor are in coplanar alignment with $|J_{RT}| = 0.5$. Comparing these two cases, which are clearly independent of the relay rotation, highlights the drastic loss caused by the misalignment which degrades both the DEB and the PTE. If instead a single solenoid relay is present (case *sol*), we see periodic improvements of DEB and PTE. These improvements follow from the relay contribution (cf. (5.4)), which bridges the misalignment between agent and anchor and hence benefits both communication and localization. The solenoid relay contribution shows the optimal improvement at roughly $\beta = 45° + z \cdot 90°$ with $z \in \mathbb{Z}$, i.e. if the relay is equally misaligned from agent and anchors. For these optimal rotations of the relay, the PTE is still lower compared to the *NoRelCopl* case. Yet, the DEB with an optimally rotated relay even exceeds that of the *NoRelCopl* case, due to the additional spatial information obtained via relay contribution. In contrast, at $\beta = z \cdot 90°$ the relay is either aligned with the anchor or approximately aligned with the agent and cannot properly mitigate their misalignment. An alignment of relay and anchor does not affect the transmit power or PTE due to a weak coupling between agent and relay, but leads to a minor relay contribution and hence minor improvement of the DEB. If the relay is however aligned with the agent, it reduces the PTE slightly but does not provide a beneficial relay contribution at the anchor. Overall, it hence even leads to a minor degradation of the DEB compared to the *NoRel* case. Lastly, we also analyze the case *3ax* for which the resonant solenoid relay is replaced with a resonant three-axis relay (cf. Fig. 2.2). Interestingly, we observe that the impact of this three-axis coil is (i) independent of its rotation and (ii) does not drastically improve the DEB or PTE. This effect is better understood if we again look at expression (5.2) for this agent-anchor pair when only a single three-axis relay is present and in case the self capacitances can be ignored:

$$\tilde{\mathbf{Z}}_\mathrm{C} = \underbrace{\begin{bmatrix} R_\mathrm{T} + j\omega L_\mathrm{T} & j\omega M_\mathrm{TR} \\ j\omega M_\mathrm{RT} & R_\mathrm{R} + j\omega L_\mathrm{R} \end{bmatrix}}_{\text{direct path}} + \underbrace{\frac{\omega^2}{R_\mathrm{Y}} \begin{bmatrix} \|\mathbf{m}_\mathrm{YT}\|^2 & \mathbf{m}_\mathrm{YT}^\mathrm{T}\mathbf{m}_\mathrm{YR} \\ \mathbf{m}_\mathrm{YR}^\mathrm{T}\mathbf{m}_\mathrm{YT} & \|\mathbf{m}_\mathrm{YR}\|^2 \end{bmatrix}}_{\text{relay path}}, \tag{5.5}$$

with $\mathbf{m}_\mathrm{YT} = [M_{\mathrm{Y_1T}}, M_{\mathrm{Y_2T}}, M_{\mathrm{Y_3T}}]^\mathrm{T}$ and $\mathbf{m}_\mathrm{YR} = [M_{\mathrm{Y_1R}}, M_{\mathrm{Y_2R}}, M_{\mathrm{Y_3R}}]^\mathrm{T}$ as respective mutual inductance vectors from the agent and anchor to all identical and uncoupled subcoils of the three-axis relay. Relying on the unilateral assumption and ideal power matching leads to

$$g^\mathrm{active} = h_\mathrm{T} = \frac{j\omega M_\mathrm{RT} + \frac{\omega^2}{R_\mathrm{Y}}\mathbf{m}_\mathrm{YR}^\mathrm{T}\mathbf{m}_\mathrm{YT}}{\sqrt{4(R_\mathrm{T} + \frac{\omega^2\|\mathbf{m}_\mathrm{YT}\|^2}{R_\mathrm{Y}})(R_\mathrm{R} + \frac{\omega^2\|\mathbf{m}_\mathrm{YR}\|^2}{R_\mathrm{Y}})}}. \tag{5.6}$$

Yet, from (2.4) we find that

$$\mathbf{m}_\mathrm{YR}^\mathrm{T}\mathbf{m}_\mathrm{YT} = \frac{K_\mathrm{YR}^\mathrm{Dip} K_\mathrm{YT}^\mathrm{Dip}}{d_\mathrm{YR}^3 d_\mathrm{YT}^3} \mathbf{o}_\mathrm{R}^\mathrm{T}\mathbf{F}_\mathrm{YR}\overbrace{\mathbf{O}_\mathrm{Y}\mathbf{O}_\mathrm{Y}^\mathrm{T}}^{\mathbf{I}_3}\mathbf{F}_\mathrm{YT}\,\mathbf{o}_\mathrm{T}\,, \tag{5.7}$$

$$\|\mathbf{m}_\mathrm{YR}\|^2 = \frac{(K_\mathrm{YR}^\mathrm{Dip})^2}{d_\mathrm{YR}^6}\mathbf{o}_\mathrm{R}^\mathrm{T}\mathbf{F}_\mathrm{YR}^2\mathbf{o}_\mathrm{R}\,, \qquad \|\mathbf{m}_\mathrm{YT}\|^2 = \frac{(K_\mathrm{YT}^\mathrm{Dip})^2}{d_\mathrm{YT}^6}\mathbf{o}_\mathrm{T}^\mathrm{T}\mathbf{F}_\mathrm{YT}^2\mathbf{o}_\mathrm{T}\,, \tag{5.8}$$

which means that the channel gain and current gain are independent of the resonant three-axis relay's orientation. Moreover, for the specific selected topology of Fig. 5.4a it further holds that $\mathbf{m}_\mathrm{YR}^\mathrm{T}\mathbf{m}_\mathrm{YT} \propto \mathbf{o}_\mathrm{R}^\mathrm{T}\mathbf{F}_\mathrm{YR}\mathbf{F}_\mathrm{YT}\,\mathbf{o}_\mathrm{T} \approx 0$, so the channel gain and current gain are only affected by the three-axis relay via their denominator and in turn only minor changes of the DEB and PTE are induced. Note that even for random topologies, the impact of a single three-axis relay remains independent of its orientation. Yet, in general $\mathbf{m}_\mathrm{YR}^\mathrm{T}\mathbf{m}_\mathrm{YT} \neq 0$ even if agent and anchor are fully misaligned, so a three-axis relay would normally benefit agent-anchor misalignments. The case $\mathbf{m}_\mathrm{YR}^\mathrm{T}\mathbf{m}_\mathrm{YT} = 0$

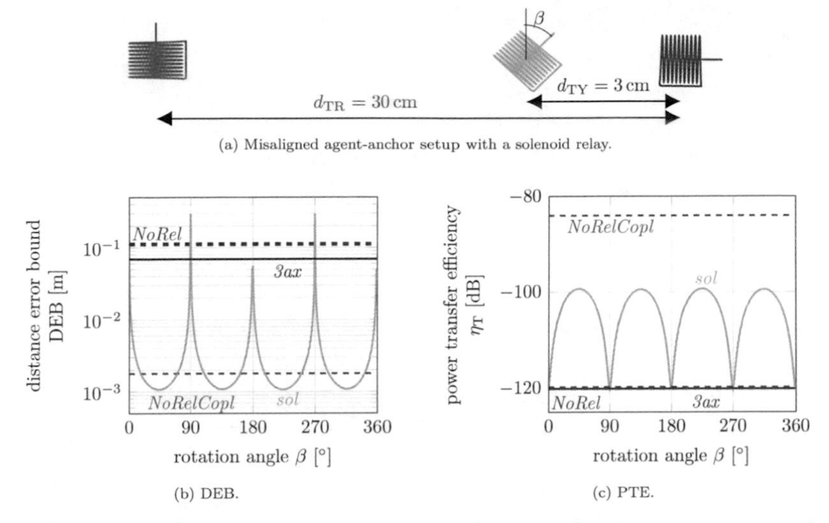

Figure 5.4: Almost fully misaligned agent-anchor pair ($J_{RT} \approx 0.01$) with a rotating passive relay inbetween. Fig. 5.4b and Fig. 5.4c show the DEB and PTE for increasing rotation angles of the relay and different cases: *NoRel* and *NoRelCopl* have no relay present but in the latter case the agent and anchor are in coplanar alignment. For *sol* and *3ax* the resonant relay uses has a solenoid antenna or a three-axis antenna, respectively. The distances and coils of Fig. 5.4a are not true to scale for better visibility.

only occurs for specific geometric structures, e.g. if the three-axis relay is placed on the line connecting agent and anchor, which leads to $\mathbf{F}_{YR} = \mathbf{F}_{YT} = \mathbf{F}_{TR}$, while agent and anchor are also misaligned.

5.2.2 Localization Impact of Multiple Resonant Passive Relays

Another interesting effect that can be created by using passive relays is that of a waveguide. Such a waveguide can be helpful if the link between agent and anchor is weak due to a large separation between the coils. In Fig. 5.5a we show a coplanar waveguide for a single agent-anchor pair, which is separated by a distance $d_{TR} = 30$ cm. The waveguide comprises a variable number N_Y of passive relays, which are all aligned identically and placed equidistantly on the line between the agent and anchor with $d^{eq} = \frac{d_{TR}}{N_Y+1}$. We distinguish between a *coplanar* and *coaxial* alignment of all involved coils and show the corresponding DEB and PTE in Fig. 5.5b and Fig. 5.5c, respectively.

For the coplanar alignment, we see that with $N_Y \leq 5$ there is almost no impact on either DEB or PTE as the passive relays are too far apart to couple well. For $5 < N_Y \leq 17$ we observe an overall decline of the DEB despite a decreasing PTE. For $N_Y > 17$ the waveguide takes full effect and the PTE increases drastically, which further reduces the DEB. Finally, at $N_Y = 29$ the DEB is almost three orders of magnitude lower compared to the case without passive relays and the PTE is roughly 40 dB higher. Additionally, we see that at $N_Y = 17$, when the PTE switches from a decreasing to an increasing behavior, there is a local maximum of the DEB. At this point the contribution of the passive relays (cf. (5.2)) overtakes the one of the direct path (not shown). Another local maximum is evident at $N_Y = 26$. For the coaxial alignment, we observe a similar behavior, however due to the stronger coupling the same effects occur for a lower number of relays and the overall performance gains are more significant. When ignoring all self capacitances for this equidistant waveguide, we obtain (cf. (2.14))

$$\mathbf{Z}_{\mathrm{C:Y}} + \mathbf{Z}_{\mathrm{Y}}^{\mathrm{load}} = \mathrm{diag}(R_{\mathrm{Y},1}, \ldots, R_{\mathrm{Y},N_{\mathrm{Y}}}) + j\omega K_{\mathrm{YY}}^{\mathrm{Dip}} J_{\mathrm{YY}} (d^{\mathrm{eq}})^{-3} \cdot \mathbf{T}\,, \tag{5.9}$$

where $\mathbf{T}$ is a real symmetric Toeplitz matrix of size $N_Y \times N_Y$ with its first row given by $[\mathbf{T}]_{1,:} = [0, 1, \frac{1}{8}, \frac{1}{27}, \ldots, \frac{1}{(N_Y-1)^3}]$. The inter-relay coupling is hence more intricate compared to the previous topologies of Sec. 5.2.1. Moreover, the relay impact on the observed input currents and possibly occuring resonance effects depend on the inverse of $\mathbf{Z}_{\mathrm{C:Y}} + \mathbf{Z}_{\mathrm{Y}}^{\mathrm{load}}$ (cf. (5.2)). For reasonably large numbers of passive relays, their impact hence cannot be described in closed form to the best of our knowledge.

In summary, the study of selected topologies shows that well-placed passive relays affect MI links on multiple levels and can be highly beneficial for localization. They can increase the PTE either by establishing a waveguide effect or by counteracting severe agent-anchor orientation misalignments. Moreover, we see that even when not increasing the PTE, well-placed passive relays are capable of providing crucial directional spatial information. However, if they get too close to the agent they trigger the *proximity problem* by detuning the impedance of the agent coil. This effect can make passive relays obsolete if it is not addressed appropriately.

5.3 Load Switching of Passive Relays

As indicated by Fig. 5.1, we want to enhance the functionality of passive relays by enabling them to switch their load, e.g. to reduce their impact on the network if they start to trigger the *proximity problem*. Moreover, with this notion of switchable passive

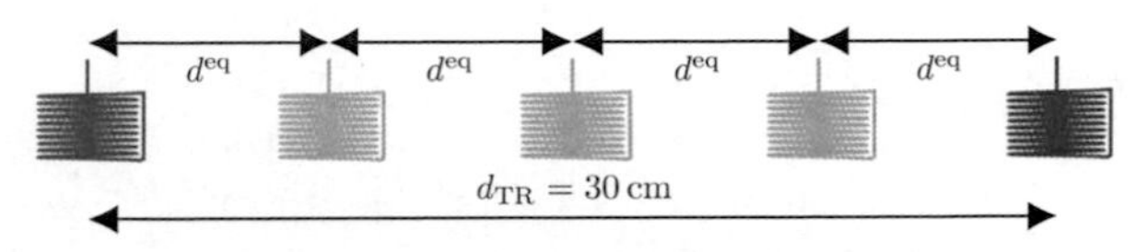

(a) Agent-anchor setup with $N_Y = 3$ relays forming an equidistant coplanar waveguide.

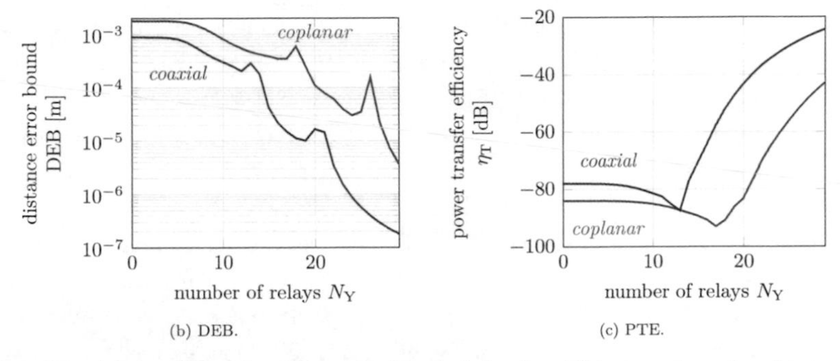

(b) DEB.

(c) PTE.

Figure 5.5: Agent-anchor pair with a well-aligned chain of equidistant passive relays inbetween. Fig. 5.4b and Fig. 5.4c show the DEB and PTE for increasing numbers N_Y of equidistant passive relays and different alignments of all coils: For *coplanar* and *coaxial* cases, all coils are placed in coplanar or coaxial alignment, respectively. The distances and coils of Fig. 5.5a are not true to scale for better visibility.

relay coils, we can generate a sequence of different current measurements at the anchors by sequentially putting different sets of passive relays into the *on* state or by adjusting their variable load capacitance. The entire process of switching different relays *on* and *off* is what we call load switching. We refer to a specific set of activated passive relays (with possibly different load values) that is used for a single measurement as load state. We further enumerate the load states used to estimate one static agent position by $k = 1, ..., N_K$, so each load state has an individual load impedance matrix $(\mathbf{Z}_{\mathrm{Y}}^{\mathrm{load}})_k$ that leads to a possibly distinct measurement of input currents. We assume that the noise of these individual observations is statistically independent, e.g. due to a sufficient time delay between measurements. As a result, the additivity of Fisher information applies and the calculation of the previously introduced bounds is straightforward according to (3.51). Passive relays which are capable of switching their load hence allow for multiple independent measurements which each provide possibly different directional information on the agents deployment. Overall, this simple concept may hence be a viable tool to amplify the beneficial localization impact of passive relays while mitigating adverse effects such as the proximity problem.

Before investigating the benefits of load switching, we distinguish between multiple empirically chosen options for relays and their switching protocols, as opposed to using no relays at all (NoRel).

All Relays (AllRel): In the simplest case, all relays are individually loaded to be resonant at the desired operating frequency during all measurement times k. For $N_K = 1$, this option coincides with our previous investigation and may easily lead to the proximity problem described above. This relay realization is purely passive, i.e. the relays do not require a variable load capacitance or a switch.

No Proximity (NoProx): This option is closely related to the *AllRel* option since most relays are simply resonantly loaded during all measurement times k. However, relays which couple strongly with the agent automatically open their load switch such that they are open-circuited and have almost no impact on the channel. In our simulative analysis, we realize this option by setting $Z_{\mathrm{Y},n_{\mathrm{Y}}}^{\mathrm{load}}$ for all passive relays n_{Y} which have a center distance of less than $2D_{\mathrm{coil}}$ to any agent coil.

Frequency Selection (FreqSel): With this option, all relays are loaded the same way as for the *AllRel* case. Yet, for the given system parametrization and design frequency f^{des}, a frequency sweep is performed and the operating frequency $\hat{f}$ which yields the lowest PEB is selected. As this approach requires knowledge of the full system constellation, it is unfeasible for practical applications. Nevertheless, it can be helpful to assess the frequency-selective behavior of the localization that may occur as a result of the passive relays. That is, the close proximity of multiple passive tags may shift the overall resonance frequency. Being able to e.g. select this new resonance frequency allows for higher channel gains and may also be beneficial for the obtained PEB.

Binary Load Switching (Binary): Another implementation considers all relays to be resonantly loaded but their switches to possibly change between the *on* and *off* state for each different measurement time k. As a result, this approach enables to conveniently and beneficially combine FIMs that are problematic if considered individually, due to having high DPEB ratio. As an example, let us consider Fig. 5.2d. We see that the directional information provided is drastically skewed to the p_y direction. However, if the relay is deactivated the directional information is heavily skewed towards the p_x direction. Combining both measurements hence leads to a more balanced directional position error ellipsoid and a significantly lower PEB. Yet, for a given relay deployment, the optimal load states in general depend on the (unknown) agent deployments. This chicken-and-egg problem of finding the optimal load states can practically be relieved if a coarse estimate of the agent deployment is available (e.g. from an initial estimate with all relays active or from other previous estimates), as the optimal load state is quite robust to small agent displacements. Yet, in this work we simply select the binary load states of all relays via genetic algorithm (GA) that is assumed to know the agent deployment and whose goal it is to minimize the PEB for a given setup. The result may hence be optimistic compared to a realistic approach which relies on a coarse initial estimate.

Variable Load Switching (Variable): The last implementation is chosen as in [28]. It realizes the variable load capacitance of any relay by combining $N_C + 1$ parallel capacitors C_{n_C} each with an individual switch $s_{n_C} \in \{0, 1\}$ and $n_C = 0, 1, \ldots, N_C$. The overall capacitance hence follows as $C_{\text{tot}} = \sum_{n_C=0}^{N_C} s_{n_C} C_{n_C}$. The first capacitors C_0 is chosen such that it enables an individual relay to realize a resonance frequency which is at most 10 % higher than the operating frequency. The remaining capacitors $C_1, \ldots, C_{N_C}$ have the same ratios as the first N_C prime numbers and the maximum load capacitance (i.e. all capacitors switches are closed, so $C_{\text{tot}} = \sum_{n_C=0}^{N_C} C_{n_C}$) yields a resonance frequency that is 10 % lower than the operating frequency. The optimization of each relay load is managed analogously to the *Binary* load switching case but clearly has a higher dimensionality. In neither case do we claim that the deployed genetic algorithms achieve optimality.

5.4 Localization Performance for Random Relay Topologies

In a next step, we compare the proposed switching patterns and examine whether they are capable of bringing the benefits of passive relays to arbitrarily arranged MI networks. To this end, we look at the three-dimensional network constellation shown in Fig. 5.6a. The agent and anchor positions are fixed, whereas the $N_{\mathrm{Y}} = 40$ passive relays are all uniformly distributed within a sphere centered at the agent, which has a 10 cm diameter. Equivalently, we can express the combination of these two parameters as relay density $\rho_{\mathrm{Y}} = 6.4\,\%$, which corresponds to the probability that a random point within the sphere also lies within the cylindrical volume of any relay. The orientation vectors of all coils are again uniformly random points on the unit sphere. For this clustered anchor scenario, we compare the different load switching algorithms with $N_K = 2$ allowed load states and $N_C = 10$ parallel capacitors for the *Variable* switching. For the subsequent analyses, we switch from using the PEB as key performance indicator to a standardized version $\overline{\mathrm{PEB}} = \sqrt{N_K}\,\mathrm{PEB}$. This standardized version is not affected by the noise averaging effect. If the load states do not change, it hence remains constant regardless of the number of used load states N_K and thus allows a fair comparison, e.g. between the *NoProx* and the *Binary* approach. The resulting empirical **C**umulative **D**istribution **F**unctions (CDFs) of this standardized PEB, the corresponding DPEB ratio, the PTE η_{T} per load state, and the receive power $\|\mathbf{y}_{\mathrm{T}}\|^2$ per load state are given in Fig. 5.6b, Fig. 5.6c, Fig. 5.6d, and Fig. 5.6e, respectively.

We find that with respect to the median performance, the case without relays yields the worst results with a median standardized PEB of about $\overline{\mathrm{PEB}} = 2\,\mathrm{cm}$. In comparison, the options that do not require a further switching of the relay loads all show a similar standardized PEB of roughly $\overline{\mathrm{PEB}} = 1.2\,\mathrm{mm}$ and hence improve the localization by almost an order of magnitude. We also observe the about 5 % to 10 % of network constellations result in outliers for the *AllRel* and *FreqSel* options. The corresponding degradation of the standardized PEB is even worse than the performance obtained with the *NoRel* option. As intended, these outliers can be successfully mitigated by deploying the *NoProx* option. Lastly, we find that both options that allow for a switching of the relay loads, namely *Binary* and *Variable*, further improve the performance and yield a median standardized PEB of approximately $\overline{\text{PEB}} = 0.35\,\text{mm}$. These options also clearly reduce the standard deviation, which indicates an increased reliability of the system. Moreover, we presume that the slight performance advantage of the *Binary* options compared to the *Variable* is the result of the latter no being able to make the passive relays resonant at the exact design frequency, due the limited number of available capacitors. Looking at the DPEB ratios in Fig. 5.6c, we observe a similar hierarchy, although the improvements are numerically not as distinct as for the standardized PEBs. However, due to the associated localization improvements for options with passive relays, we assume that observed asymmetry of the DPEBs in these cases often stems from *too much* information in one direction, as opposed to having *too little* information in another direction. Moreover, examining the PTE of the channel in Fig. 5.6d, we find that with exception of the *FreqSel* option the relays do not have any relevant impact on this quantity. The decrease of the PTE for the *FreqSel* option is a result of the impaired power matching networks on the anchor side at frequencies other than the design frequency. Overall, the plot highlights that the contribution of all passive tags on the transimpedance matrix does not increase the PTE in a statistically significant manner. Moreover, the operating frequencies which minimize the PEB, do not coincide with those operating frequencies which maximize the PTE between agent and anchors or the receive power. For the *FreqSel* option, this is further emphasized by Fig. 5.7, which shows the empirically estimated PDFs of the selected frequency which either minimizes the PEB, maximizes the PTE or maximizes the received sum power. While the latter two quantities exhibit a symmetric PDF centered at the design frequency, the PDF for the frequency that minimize the PEB is shifted to the lower frequencies. Moreover, the selected frequency for the minimal PEB only coincided with the other selected frequencies for the PTE and receive power in 7 % of all random realizations (not shown). Lastly, examining Fig. 5.6e, we find that the receive power is negatively affected for the *FreqSel* option for the above stated reasons and also for the *AllRel* option as a result of the known proximity problem. The other options however consistently mitigate the proximity problem to yield a low PEB while maintaining approximately the same receive sum power as the option without any relays.

The previous investigation showed that even for random network constellations, passive relays do improve the localization capabilities of a system if many of them are in close vicinity to the agent. The extent of this improvement can be further influenced drastically with well-suited load states. However, their beneficial impact

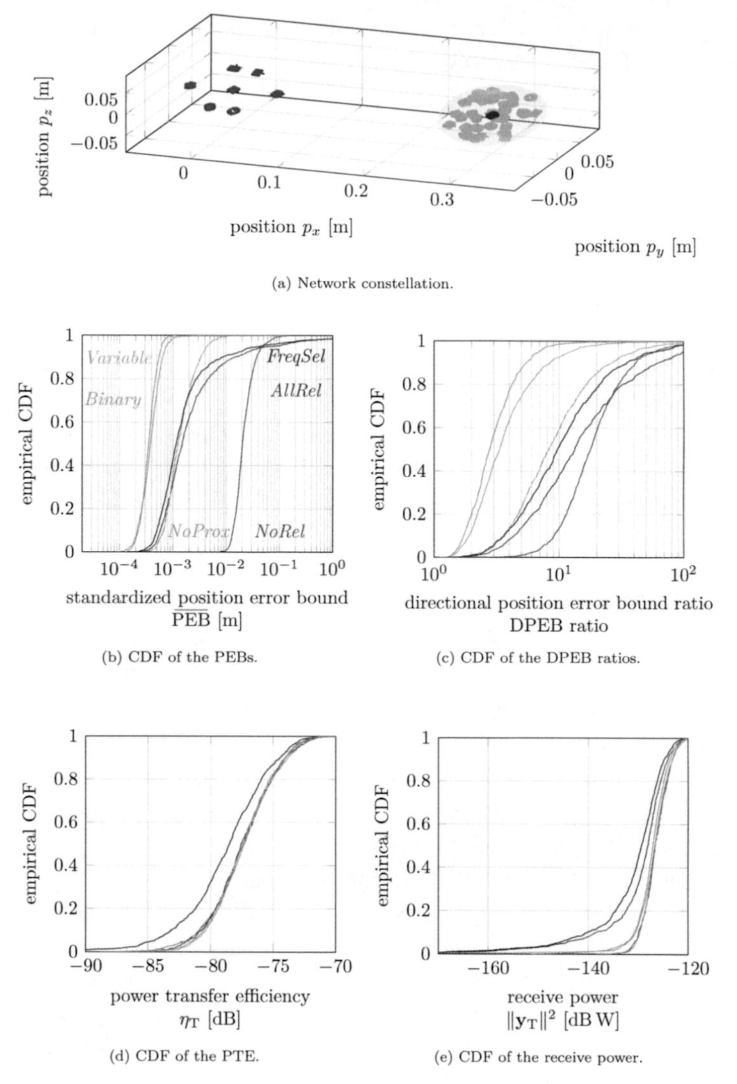

(a) Network constellation.

(b) CDF of the PEBs.

(c) CDF of the DPEB ratios.

(d) CDF of the PTE.

(e) CDF of the receive power.

Figure 5.6: Statistical comparison of different load switching algorithms for a three-dimensional clustered anchor scenario with $N_{\text{Y}} = 40$ passive relays randomly distributed close to the agent. All coil orientations are uniformly random points on the unit sphere.

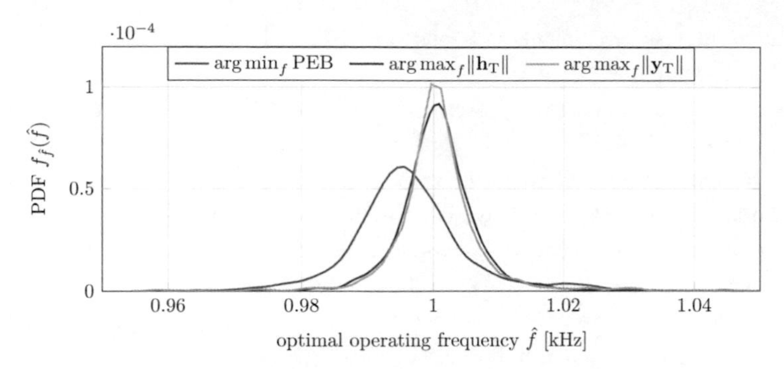

Figure 5.7: Probability density estimates of the selected operating frequency when deploying the *FreqSel* option for the network constellation of Fig. 5.6a with different figures of merit.

requires knowledge of each relay's deployment vector, which can be hard to obtain in a practical setting. In a next step, we hence study how the beneficial impact of the relays depends on the number of relays and load states. To this end, Fig. 5.8a shows the standardized PEB for the previous network constellation with a changing number of relays or equivalently, a varying relay density. To improve the visibility, we omit the results of the *FreqSel* and *Variable* options, which are almost identical to those of the *AllRel* and *Binary* options, respectively. We find that for $N_\mathrm{Y} = 1$ relay, the different options yield comparable results, which implicates that a single relay is almost always activated as it is unlikely in the agent's direct proximity. For more relays, sophisticated switching of the load states via *Binary* option consistently leads to an improved median standardized PEB as well as an improved interdecile range. That is, the median standardized PEB appears to roughly scale with $\overline{\mathrm{PEB}} \propto N_\mathrm{Y}^{-1}$. In comparison, the *AllRel* and *NoProx* options scale approximately with $\overline{\mathrm{PEB}} \propto N_\mathrm{Y}^{-\frac{1}{2}}$ but show diminishing returns for an increasing number of relays and fall almost an order of magnitude short at $N_\mathrm{Y} = 100$. Yet, while the median performance between the *AllRel* and *NoProx* options only show minuscule differences, we find that the *NoProx* options offers a severe robustness gain as indicated by the difference of the interdecile ranges. Also note that from Sec. 4.3 we saw that for a distributed anchor scenario the number of anchors only affected the PEB according to PEB $\propto \frac{1}{N_\mathrm{R}^2}$ via noise averaging effect. This effect occurred after a certain threshold of anchors was met and new anchors were unable to obtain additional directional or spatial information. We suspect that this behavior also holds for a clustered anchor scenario, where additional anchors cannot compensate the lack of directional information in specific directions. For these scenarios, using a large number of randomly deployed passive relays shows to be a simple way to improve the localization.

Next, we also visualize the impact of the number of load states N_K in dependence of the number of relays N_Y for the same network constellation via intensity plots in Fig. 5.8b and Fig. 5.8c. More precisely, the intensity plots show the resulting median standardized PEB and median DPEB ratio for different parameter choices when the *Binary* option is used. The second row of Fig. 5.8b is hence identical to the solid green line of Fig. 5.8a. Surprisingly, we observe that the number of load states plays a subordinate role compared to the number of relays. Moreover, while allowing for $N_K = 2$ load states leads to minor improvements, any further increase does not significantly impact the median standardized PEB or the median DPEB ratio. We assume that this behavior is a result of the random network constellation. More specifically, the example of Fig. 5.2d, which was mentioned when the *Binary* option was introduced, is unlikely to occur for random network deployments. So while there theoretically are constellations for which additional load states are highly beneficial, the chance of those constellations actually occurring in random constellations is evidently low. As a result, load-switching over multiple load states may not always be a worthwhile endeavor, especially when considering the additional increase in hardware complexity that comes with this operation. As demonstrated in [65], we hence see its main benefit as providing multiple independent observations if only an insufficient number of anchors is located in range of the agent.

5.4.1 Range Extension with Randomly Placed Passive Relays

So far, we have used the PEB as performance indicator since it reflects the potential localization accuracy. Yet, we have not considered the overall supportable range of a system, which is another important performance indicator for MI systems in practice. That is, it may not be necessary to decrease the PEB by orders of magnitude for a given distance, but rather to consistently uphold a certain PEB and making sure that it does not surpass a given threshold, even if a sensors moves further away from the anchors. In Fig. 5.9a we show a coplanar setup similar to that of Fig. 5.6a but now for increasing distances d_TR between the agent and the centrally located anchor. Additionally, we place $N_\mathrm{Y} = 100$ passive relays uniformly random within the depicted rectangle and in a coplanar alignment. The corresponding standardized PEB, DPEB ratio, and PTE are shown for selected load switching schemes in Fig. 5.9b, Fig. 5.9c, and Fig. 5.9d, respectively. It is evident that the beneficial impact of the passive relays on the PEB

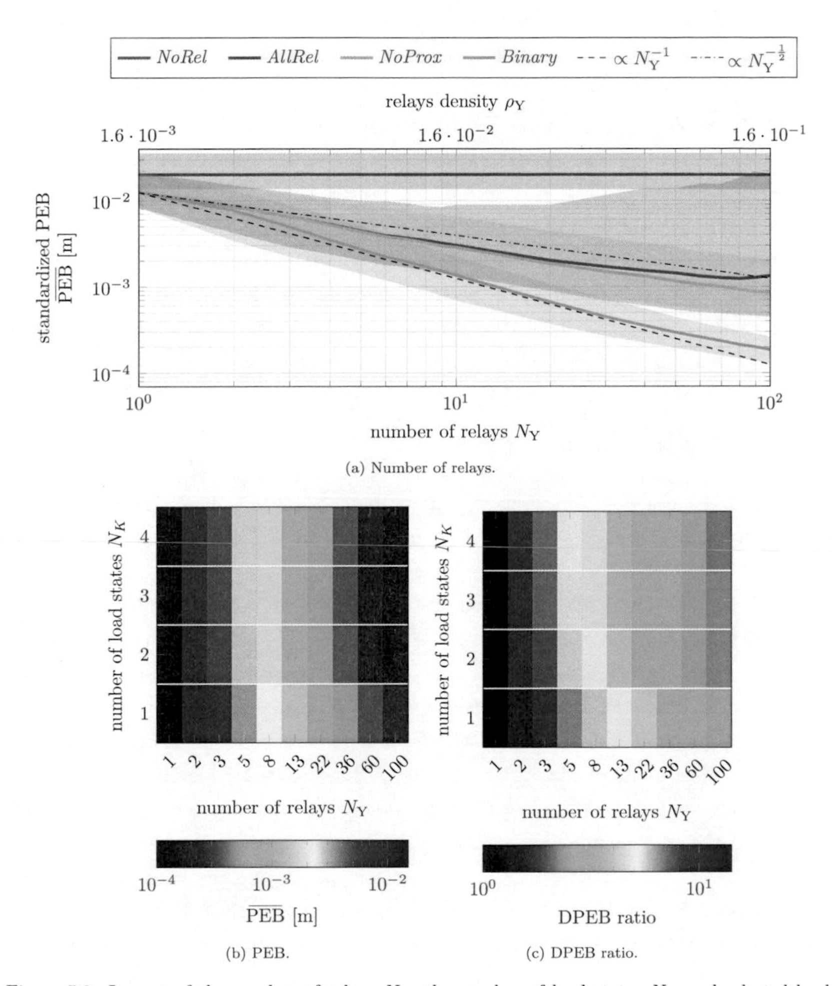

(a) Number of relays.

(b) PEB.

(c) DPEB ratio.

Figure 5.8: Impact of the number of relays N_Y, the number of load states N_K and selected load-switching options on the standardized PEB. The network constellation is chosen randomly as in Fig. 5.6a.

increases over distance both for the *NoProx* and *Binary* scheme. In constrast, the PTE is degraded for both schemes, i.e. no waveguide is established. Regarding the DPEB ratio, we find that without relays there is a steep decrease when the agent leaves the immediate vicinity of the anchors, followed by a continuous increase of the DPEB ratio, which is consistent with our observations from Sec. 4.4. In comparison, the *NoProx* and *Binary* options quickly accomplish to decrease the median DPEB ratio to roughly DPEB ratio ≈ 5 and DPEB ratio ≈ 2, respectively. For the *Binary* scheme, this DPEB ratio is upheld as long as the agent is placed within the coverage area of the relays. Moreover, while the *NoProx* does not manage to fully upheld the reduced DPEB ratio, it drastically slows its increase with distance. For a PEB $= 1\,\mathrm{cm}$ threshold, these effects allow for a range extension from $30\,\mathrm{cm}$ to either roughly $50\,\mathrm{cm}$ (*NoProx*) or roughly $90\,\mathrm{cm}$ (*Binary*). When the agent leaves the field of relays, the impact quickly subsides and at $d_{\mathrm{TR}} = 1.1\,\mathrm{m}$ no more beneficial impact can be observed.

5.5 Conclusions

We studied the use of passive relays (resonantly loaded coil antennas) as auxiliary nodes to improve the accuracy, range and robustness of MI localization. In this process, we showed that a passive relay provides an additional signal path from an agent to an anchor. Via this additional contribution, a well-placed passive relay can provide additional directional information about the agent position. As a result, a single passive relay is capable of decreasing the PEB by more than an order of magnitude for clustered anchor constellations. Moreover, for selected placements we demonstrated that passive relays can also increase the PTE from the agent to anchors either by bridging misalignments of the coils or by establishing a waveguide effect. Yet, if the passive relays couple too strongly with agents or anchors, they trigger the same mutual detuning of the coil impedances that was already observed for agent-anchor pairs in Sec. 4.1 and can deteriorate the PTE, the transmit power, and the PEB. This adverse *proximity problem* can be mitigated by extending the functionality of the passive relays and enabling them to switch between different loads. With this extension, we further demonstrated that the improved localization accuracy and range offered by well-placed passive relays can also be obtained via arbitrarily-placed swarms, if a sufficient number of passive relays is present.

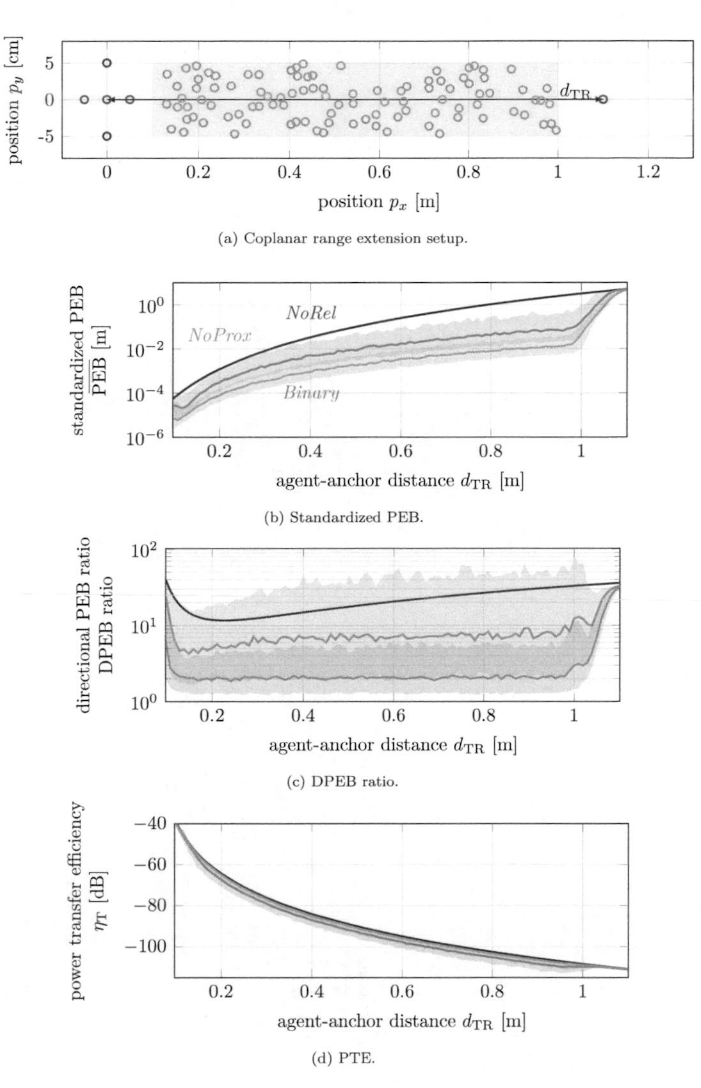

(a) Coplanar range extension setup.

(b) Standardized PEB.

(c) DPEB ratio.

(d) PTE.

Figure 5.9: Range Extension study for a coplanar setup with uniformly random placement of $N_Y = 100$ passive relays. The standardized PEB, the DPEB ratio and the PTE are shown for increasing agent-anchor distances d_{TR} and selected load switching schemes.

Chapter 6

Cooperative Localization

Note: Parts of this chapter have been published by us in [92]. This work hence exhibits similarities regarding formulations and visualizations.

As demonstrated in Cpt. 5, passive relays are a powerful means to mitigate inherent issues of MI networks and thus improve the localization accuracy and range. Yet, for them to be useful they need to have a sufficient coupling with the agents, which either requires them to be in close proximity or for the coils to have high quality factors. It is furthermore imperative to know the relay deployment, which may be unfeasible for some practical settings. In this chapter, we will hence investigate cooperation of agent nodes as an alternative means to deploying passive relays. Throughout this chapter, we again assume to fully know the deployment of the anchors, whereas the deployment of all agents is unknown. Further, we now consider sensor nodes which comprise either a solenoid coil or a three-axis coil (cf. Sec. 2.2).

In non-cooperative localization networks, the agents are typically active nodes that independently transmit a pilot signal to all anchors. The anchors then forward all received signals to a central unit, where they are combined in order to localize each agent individually. However, for our definition of cooperation we rely on a different working principle, which requires a full mesh of all the nodes in the network. That is, we assume that an agent also has the capability to receive the pilot signals of other agents, extracts relevant information from these signals, and then forwards this information (added with the corresponding agent ID) to an anchor during its next own communication session. This would ultimately lead to the central unit having information on all inter-node signals. Compared to the traditional non-cooperative scenario, the localization does thus not only rely on the knowledge of the agent-anchor channels, but can also incorporate the information of all agent-agent channels. Alternatively, cooperating agents may thus also be interpreted as anchors with an unknown deployment. This concept originates from traditional radio and has been investigated thoroughly, e.g. for UWB technology (cf. [126–128]) or 5G networks (cf. [129]), where an improved localization accuracy and robustness were identified as key benefits. While cooperative localization has not been studied for MI networks so far, we expect it to be particularly advantageous as it (i) generally increases the number of independent observations, (ii) can exploit short inter-agent distances that are associated with a strong channel gain, (iii) counteract misalignments of agent-anchor observations, and (iv) may generally provide further directional information to balance the DPEBs via inter-agent links.

6.1 Weakly-Coupled Channel Gain and Noise Model

In the course of this chapter, we will use $m = 1, \dots, M$ as index that is exclusively used to denote the agents and $n = 1, \dots, M, \dots, M + N$ as index that describes both agents and anchors. This distinction is an important change compared to the notation used in previous chapters. Moreover, as we will limit the analyses to strictly active agent nodes without the presence of passive relays, we do not require the nodes to be strongly coupled or in close proximity for localization purposes. We hence apply a minimum distance constraint between any pair of nodes that allows us to simplify the more complex system model from Sec. 3.1. That is, we assume all links (even agent-agent or anchor-anchor links) to be weakly coupled in conjunction with the unilateral assumption, ideal power matching and the magnetic dipole approximation for all coils (cf. Sec. 3.3.3). As a result, the channel gain for each SISO coil pair m, n is given by (cf. (2.3) and (3.45))

$$h_{\mathrm{T}} = h_{m,n} = \frac{j\omega M_{m,n}}{\sqrt{4R_m R_n}} = \frac{j a_{m,n}}{d_{m,n}^3}\ \mathbf{o}_m^{\mathrm{T}} \left(\frac{3}{2}\mathbf{u}_{m,n}\mathbf{u}_{m,n}^{\mathrm{T}} - \frac{1}{2}\mathbf{I}_3\right) \mathbf{o}_n\,, \tag{6.1}$$

with $a_{m,n} = \frac{\omega K^{\mathrm{Dip}}}{\sqrt{4R_m R_n}}$. We can further extend this complex scalar channel gain to the full channel matrix $\mathbf{H}_{m,n} \in \mathbb{C}^{3\times 3}$ between a three-axis coil pair via (2.4), which leads to

$$\mathbf{H}_{m,n} = \frac{j a_{m,n}}{d_{m,n}^3}\ \mathbf{O}_m^{\mathrm{T}} \underbrace{\left(\frac{3}{2}\mathbf{u}_{m,n}\mathbf{u}_{m,n}^{\mathrm{T}} - \frac{1}{2}\mathbf{I}_3\right)}_{\mathbf{F}_{m,n}} \mathbf{O}_n\,. \tag{6.2}$$

Note that the orientation matrices $\mathbf{O}_m$ and $\mathbf{O}_n$ coincide with the rotation matrix definition of (A.18) and can hence be fully expressed by the three Euler angles of the corresponding coil.

The measured received signal $y_{m,n}^{\mathrm{meas}}$ comprising the thermal noise and LNA noise then follows as

$$y_{m,n}^{\text{meas}} = h_{m,n}\, x_{m,n} + w_{m,n} \overset{*}{=} h_{m,n}^{\text{meas}}\,, \tag{6.3}$$

with $w_{m,n} \overset{\text{i.i.d}}{\sim} \mathcal{CN}(0, \sigma_{\text{sol}}^2)$ and $\sigma_{\text{sol}}^2 = k_{\text{B}}TB + \frac{(\sigma^{\text{LNA}})^2}{4}\left(\frac{(R^{\text{LNA}})^2}{R^{\text{ref}}} + R^{\text{ref}} - 2R^{\text{LNA}}\text{Re}(\rho^{\text{LNA}})\right)$. This signal-based representation is equivalent to our previous current-based model up to a scaling factor of $\sqrt{R^{\text{ref}}}$. Moreover, as we assume ideal power matching the transmit signal $x_{m,n}$ does not depend on the network constellation and we can choose $x_{m,n} = 1$ for mathematical convenience, which leads to equality in second part of (6.3). As a result, the receive signal is equal to the noisy channel gain $h_{m,n}^{\text{meas}}$ of the corresponding coil pair. Next, if we operate each transmitting subcoil of each three-axis coil in a time multiplexed fashion, the analogous representation for the noisy MIMO observations is given by

$$\mathbf{H}_{m,n}^{\text{meas}} = \mathbf{H}_{m,n} + \mathbf{W}_{m,n}\,, \tag{6.4}$$

with $[\mathbf{W}_{m,n}]_{k,l} \overset{\text{i.i.d}}{\sim} \mathcal{CN}(0, \sigma_{3\text{ax}}^2)$ where k, l are the indices of the corresponding subcoils. Lastly, we choose $\sigma_{3\text{ax}}^2 = 3\sigma_{\text{sol}}^2$ to counterbalance the difference in transmit power between a solenoid and a three-axis coil.

For our proposed application, we collect a multitude of these noisy channel gains or channel matrices in a set and combine them to estimate all agent positions. For the non-cooperative scenario, we only obtain channel measurements that correspond to agent-anchor links, whereas for the cooperative case we additionally have the agent-agent channel measurements, i.e.

$$\mathcal{H}_{3\text{ax}}^{\text{NonCoop}} = \left\{\mathbf{H}_{m,n}^{\text{meas}} \mid m \in \{1, \ldots, M\}, n \in \{M+1, \ldots, M+N\}\right\}, \tag{6.5}$$

$$\mathcal{H}_{3\text{ax}}^{\text{Coop}} = \left\{\mathbf{H}_{m,n}^{\text{meas}} \mid m \in \{1, \ldots, M\}, n \in \{1, \ldots, M+N\} \setminus \{m\}\right\}, \tag{6.6}$$

or equivalently when considering a setup that exclusively uses solenoid coils

$$\mathcal{H}_{\text{sol}}^{\text{NonCoop}} = \left\{h_{m,n}^{\text{meas}} \mid m \in \{1, \ldots, M\}, n \in \{M+1, \ldots, M+N\}\right\}, \tag{6.7}$$

$$\mathcal{H}_{\text{sol}}^{\text{Coop}} = \left\{h_{m,n}^{\text{meas}} \mid m \in \{1, \ldots, M\}, n \in \{1, \ldots, M+N\} \setminus \{m\}\right\}. \tag{6.8}$$

As described in Sec. 2.2, the full placement and orientation of each coil n is given by the six-dimensional deployment vector $\boldsymbol{\psi}_n$. However, we established via (2.3) that the last orientation parameter γ around a solenoid's cylindrical main axis has almost no impact on the measurements and its estimation was not necessary. For three-axis coils this is not the case as the rotation around any axis affects all of its solenoid subcoils jointly and has a relevant impact on the measurement. To ensure that this distinction is clear, we use different subscripts sol and 3ax, if the affiliation of any variables is not directly clear from the setting. That is, for the deployment vectors we e.g. distinguish between $(\boldsymbol{\psi}_m)_{\text{sol}} = [\mathbf{p}_m^{\mathrm{T}}, \alpha_m, \beta_m]^{\mathrm{T}}$ for the solenoid case and $(\boldsymbol{\psi}_m)_{\text{3ax}} = [\mathbf{p}_m^{\mathrm{T}}, \alpha_m, \beta_m, \gamma_m]^{\mathrm{T}}$ for the three-axis case.

6.2 Position Error Bounds

In order to compare cooperative and non-cooperative localization for MI systems, we will adjust the earlier definition of the PEB from Sec. 3.4 in this section. Since all nodes are ideally power matched, it follows that $\mathbf{K} \propto \mathbf{I}_N$ for the non-cooperative case and $\mathbf{K} \propto \mathbf{I}_{M+N-1}$ for the cooperative case (cf. (3.22)). As a result, the elements of the cooperative FIM follow from (3.49) and (3.51) as summation over all available channel gain derivatives, e.g. for solenoid coils as

$$[\boldsymbol{\mathcal{I}}^{\text{Coop}}]_{i,j} = \frac{2}{\sigma_{\text{sol}}^2} \mathrm{Re}\left(\sum_{m=1}^{M} \sum_{\substack{n=1 \\ n \neq m}}^{M+N} \frac{\partial h_{m,n}^{\mathrm{H}}}{\partial [\boldsymbol{\Psi}]_i} \frac{\partial h_{m,n}}{\partial [\boldsymbol{\Psi}]_j} \right). \tag{6.9}$$

However, for any deployment variable $[\boldsymbol{\Psi}]_i$ which does not belong to the agent m, it is clear that $\frac{\partial h_{m,n}^{\mathrm{H}}}{\partial [\boldsymbol{\Psi}]_i} = 0$. Limiting the derivatives to $\frac{\partial h_{m,n}}{\partial [\boldsymbol{\psi}_m]_i}$ and $\frac{\partial h_{m,n}}{\partial [\boldsymbol{\psi}_m]_j}$ instead hence allows us to drop the first summation (cf. (6.11) and (6.12)). Moreover, if $[\boldsymbol{\Psi}]_i$ belongs to agent m and $[\boldsymbol{\Psi}]_j$ belongs to a different agent $n \in \{1, \dots, M\} \setminus \{m\}$, then there is only one single channel gain which yields non-zero results, namely the one between agent m and agent n, i.e. $\mathrm{Re}\left(\frac{\partial h_{m,n}^{\mathrm{H}}}{\partial [\boldsymbol{\psi}_m]_i} \frac{\partial h_{m,n}}{\partial [\boldsymbol{\psi}_n]_j} \right) \neq 0$ (cf. (6.13)). Thus, only considering non-zero blocks of $\boldsymbol{\mathcal{I}}^{\text{Coop}}$, we employ the established block matrix form [118, 128]

$$\boldsymbol{\mathcal{I}}^{\text{Coop}} = \begin{bmatrix} \mathbf{N}_1 + \mathbf{C}_{1,1} & \mathbf{C}_{1,2} & \cdots & \mathbf{C}_{1,M} \\ \mathbf{C}_{2,1} & \mathbf{N}_2 + \mathbf{C}_{2,2} & \cdots & \mathbf{C}_{2,M} \\ \vdots & \vdots & \ddots & \vdots \\ \mathbf{C}_{M,1} & \mathbf{C}_{M,2} & \cdots & \mathbf{N}_M + \mathbf{C}_{M,M} \end{bmatrix}, \tag{6.10}$$

where the elements of each individual 5×5 submatrix for the solenoid case are found as (cf. [115], (3.50))

$$([\mathbf{N}_m]_{i,j})_{\text{sol}} = \frac{2}{\sigma^2_{\text{sol}}} \sum_{n=M+1}^{M+N} \text{Re}\left(\frac{\partial h^{\text{H}}_{m,n}}{\partial[\boldsymbol{\psi}_m]_i} \frac{\partial h_{m,n}}{\partial[\boldsymbol{\psi}_m]_j} \right), \qquad \underset{m}{\text{agent}} \rightarrow \underset{M+1\leq n\leq M+N}{\text{anchor}}, \tag{6.11}$$

$$([\mathbf{C}_{m,m}]_{i,j})_{\text{sol}} = \frac{2}{\sigma^2_{\text{sol}}} \sum_{\substack{n=1\\n\neq m}}^{M} \text{Re}\left(\frac{\partial h^{\text{H}}_{m,n}}{\partial[\boldsymbol{\psi}_m]_i} \frac{\partial h_{m,n}}{\partial[\boldsymbol{\psi}_m]_j} \right), \qquad \underset{m}{\text{agent}} \rightarrow \underset{1\leq n\leq M}{\text{agents}}, \tag{6.12}$$

$$([\mathbf{C}_{m,n}]_{i,j})_{\text{sol}} = \frac{2}{\sigma^2_{\text{sol}}} \text{Re}\left(\frac{\partial h^{\text{H}}_{m,n}}{\partial[\boldsymbol{\psi}_m]_i} \frac{\partial h_{m,n}}{\partial[\boldsymbol{\psi}_n]_j} \right)\Bigg|_{n\in\{1,\dots,M\}\setminus\{m\}}, \qquad \underset{m}{\text{agent}} \rightarrow \underset{n}{\text{agent}}, \tag{6.13}$$

with $i, j \in \{1, \dots, 5\}$. The diagonal blocks $\mathbf{N}_m$ account for the measurements from agent m to all anchors, whereas the diagonal blocks $\mathbf{C}_{m,m}$ account for all inter-agent measurements of this agent, in case the other agents act as anchors. However, as their deployment is unknown, this uncertainty in their function as anchor is expressed by the off-diagonal blocks $\mathbf{C}_{m,n}$. For three-axis coils the FIM has the same structure as in (6.10) and the scalar elements of the individual 6×6 sub matrices follow analogously as

$$([\mathbf{N}_m]_{i,j})_{\text{3ax}} = \frac{2}{\sigma^2_{\text{3ax}}} \sum_{n=M+1}^{M+N} \text{tr}\left(\frac{\partial \mathbf{H}^{\text{H}}_{m,n}}{\partial[\boldsymbol{\psi}_m]_i} \frac{\partial \mathbf{H}_{m,n}}{\partial[\boldsymbol{\psi}_m]_j} \right), \tag{6.14}$$

$$([\mathbf{C}_{m,m}]_{i,j})_{\text{3ax}} = \frac{2}{\sigma^2_{\text{3ax}}} \sum_{\substack{n=1\\n\neq m}}^{M} \text{tr}\left(\frac{\partial \mathbf{H}^{\text{H}}_{m,n}}{\partial[\boldsymbol{\psi}_m]_i} \frac{\partial \mathbf{H}_{m,n}}{\partial[\boldsymbol{\psi}_m]_j} \right), \tag{6.15}$$

$$([\mathbf{C}_{m,n}]_{i,j})_{\text{3ax}} = \frac{2}{\sigma^2_{\text{3ax}}} \text{tr}\left(\frac{\partial \mathbf{H}^{\text{H}}_{m,n}}{\partial[\boldsymbol{\psi}_m]_i} \frac{\partial \mathbf{H}_{m,n}}{\partial[\boldsymbol{\psi}_n]_j} \right)\Bigg|_{n\in\{1\dots M\}\setminus\{m\}}, \tag{6.16}$$

with $i, j \in \{1, \dots, 6\}$ and where we used $\text{vec}\left(\frac{\partial \mathbf{H}_{m,n}}{\partial[\boldsymbol{\psi}_m]_i}\right)^{\text{H}} \text{vec}\left(\frac{\partial \mathbf{H}_{m,n}}{\partial[\boldsymbol{\psi}_m]_j}\right) = \text{tr}\left(\frac{\partial \mathbf{H}^{\text{H}}_{m,n}}{\partial[\boldsymbol{\psi}_m]_i} \frac{\partial \mathbf{H}_{m,n}}{\partial[\boldsymbol{\psi}_m]_j}\right)$. The required derivatives of the channel matrices are a straight-forward extension from the scalar case.

Consequently, in the cooperative case, the PEB of the first agent ($m = 1$) is given by

$$\text{PEB}_1^{\text{Coop}} = \sqrt{\text{tr}\left(\left[\left(\boldsymbol{\mathcal{I}}^{\text{Coop}}\right)^{-1}\right]_{1:3,1:3}\right)}. \tag{6.17}$$

By adjusting the indexing, the cooperative PEB of any other agent m can be calculated analogously. However, the PEB of any such agent would be statistically identical to that of agent 1, since all agent deployment vectors are i.i.d. random vectors. Our specific choice does therefore not affect the generality.

Lastly, for the non-cooperative case it holds that $\mathbf{C}_{m,n} = \mathbf{0} = \mathbf{C}_{m,m}$, so $(\boldsymbol{\mathcal{I}}^{\mathrm{NonCoop}})^{-1} = \mathrm{diag}(\mathbf{N}_1^{-1}, \ldots, \mathbf{N}_m^{-1})$ and the PEB of each agent only relies on its corresponding non-cooperative diagonal block, i.e.

$$\mathrm{PEB}_1^{\mathrm{NonCoop}} = \sqrt{\mathrm{tr}\left(\left[\mathbf{N}_1^{-1}\right]_{1:3,1:3}\right)}. \tag{6.18}$$

6.3 Cooperative and Non-Cooperative Deployment Estimators

Compared to the studies of Cpt. 4 and Cpt. 5, we will also provide three different practical approaches to obtain estimates of the agents' deployment vectors for either solenoid or three-axis coils. These approaches are based on (i) a numerical **L**east-**S**quares (LS) estimation, (ii) a pairwise ML distance estimation with subsequent position estimation, and (iii) a combination of both.

6.3.1 LS Estimation

Having a statistical model for our observations (cf. (6.4)), the straightforward approach to estimate all agent deployments would be via joint ML estimator, which in our Gaussian case coincides with the joint LS estimator. For a cooperative system, this is given by the $5M$-dimensional or $6M$-dimensional nonlinear LS problems

$$(\hat{\boldsymbol{\psi}}_1, \ldots, \hat{\boldsymbol{\psi}}_M)_{\mathrm{3ax}}^{\mathrm{Coop}} = \underset{(\boldsymbol{\psi}_1,\ldots,\boldsymbol{\psi}_M)_{\mathrm{3ax}}}{\arg\min} \sum_{m=1}^{M} \sum_{\substack{n=1\\ n\neq m}}^{M+N} \left\| \mathbf{H}_{m,n}^{\mathrm{meas}} - \mathbf{H}_{m,n} \right\|_{\mathrm{F}}^2, \tag{6.19}$$

$$(\hat{\boldsymbol{\psi}}_1, \ldots, \hat{\boldsymbol{\psi}}_M)_{\mathrm{sol}}^{\mathrm{Coop}} = \underset{(\boldsymbol{\psi}_1,\ldots,\boldsymbol{\psi}_M)_{\mathrm{sol}}}{\arg\min} \sum_{m=1}^{M} \sum_{\substack{n=1\\ n\neq m}}^{M+N} \left| h_{m,n}^{\mathrm{meas}} - h_{m,n} \right|^2, \tag{6.20}$$

where $\mathbf{H}_{m,n}$ and $h_{m,n}$ are functions of $\boldsymbol{\psi}_m$ and $\boldsymbol{\psi}_n$ if $n \in \{1, \ldots, M\}$ (agent-agent link) or functions of just $\boldsymbol{\psi}_m$ if $n > M$ (agent-anchor link with known anchor parameters). For the non-cooperative scenario, the problem can be decomposed into M individual five-dimensional or six-dimensional nonlinear LS problems:

$$(\hat{\psi}_m)_{3\mathrm{ax}}^{\mathrm{NonCoop}} = \underset{(\psi_m)_{3\mathrm{ax}}}{\arg\min} \sum_{n=M+1}^{M+N} \left\| \mathbf{H}_{m,n}^{\mathrm{meas}} - \mathbf{H}_{m,n} \right\|_{\mathrm{F}}^2, \tag{6.21}$$

$$(\hat{\psi}_m)_{\mathrm{sol}}^{\mathrm{NonCoop}} = \underset{(\psi_m)_{\mathrm{sol}}}{\arg\min} \sum_{n=M+1}^{M+N} \left| h_{m,n}^{\mathrm{meas}} - h_{m,n} \right|^2. \tag{6.22}$$

To the best of our knowledge, these estimation rules can only be realized via numerical solvers. This approach is however computationally expensive and might easily converge to local minima as shown by [61] and later visualized in Fig. 6.4. The inherent advantage that is gained by using cooperation might hence be directly nullified due to the additional challenges in numerical optimization.

6.3.2 Pairwise Estimation for Non-Cooperative Localization

In order to alleviate the computational complexity issue we propose a second approach, which is based on the individual ML estimation for the pairwise links between agents and anchors. For a pair of solenoid coils and under the assumption that one coil's orientation and the direction of departure are uniformly randomly distributed points on the unit sphere, the ML distance estimator was already stated by [130] as

$$(\hat{d}_{m,n}^{\mathrm{ML}})_{\mathrm{sol}} = \sqrt[3]{\frac{1}{2} \cdot \frac{a_{m,n}}{\left| h_{m,n}^{\mathrm{meas}} \right|^2}}. \tag{6.23}$$

An established approach to combine multiple pairwise agent-anchor ($n > M$) distance estimates to a position estimate is the so called multilateration (cf. [131]), i.e. solving the three-dimensional LS problem

$$(\hat{\mathbf{p}}_m)_{\mathrm{sol}} = \underset{\mathbf{p}_m}{\arg\min} \sum_{n=M+1}^{M+N} k_{m,n} \left| (\hat{d}_{m,n}^{\mathrm{ML}})_{\mathrm{sol}} - d_{m,n}(\mathbf{p}_m) \right|^2, \tag{6.24}$$

individually for each agent. The weights for each agent are chosen as $k_{m,n} = 0.25$ for the four strongest anchor observations (identified by $\left| h_{m,n}^{\mathrm{meas}} \right|$) and $k_{m,n} = 0$ for all other links. Using a weighting for a multilateration is common practice to ignore drastic outliers which otherwise impair the overall estimation quality (cf. [132]). In our case the simple weighting is used to mitigate the over-estimation of distances that are the result of drastically misaligned coil pairs.

For the three-axis coil case, there is is no known ML distance estimator. We will instead derive it in the next steps. To this end, we consider the Gaussian log-**L**ikeli**H**ood **F**unction (LHF) $L(\boldsymbol{\psi}_m)$ of an agent's deployment vector based on one measured pair-wise agent-anchor[1] channel matrix $\mathbf{H}_{m,n}^{\text{meas}}$ with $M < n \leq M + N$. We discard the constant summand from $L(\boldsymbol{\psi}_m)$ for mathematical convenience and find the resulting shifted log-LHF $\tilde{L}(\boldsymbol{\psi}_m) = L(\boldsymbol{\psi}_m) + \log(\pi^9 \sigma_{3\text{ax}}^{18})$ as

$$\begin{aligned}
\tilde{L}(\boldsymbol{\psi}_m) &= -\frac{1}{\sigma^2} \operatorname{tr}\left(\left(\mathbf{H}_{m,n}^{\text{meas}} - \mathbf{H}_{m,n}\right)^{\text{H}} \left(\mathbf{H}_{m,n}^{\text{meas}} - \mathbf{H}_{m,n}\right)\right), \\
&\propto -\operatorname{tr}\left(\left(\mathbf{H}_{m,n}^{\text{meas}}\right)^{\text{H}} \mathbf{H}_{m,n}^{\text{meas}}\right) - \operatorname{tr}\left((\mathbf{H}_{m,n})^{\text{H}} \mathbf{H}_{m,n}\right) + 2\text{Re}\left(\operatorname{tr}\left(\left(\mathbf{H}_{m,n}^{\text{meas}}\right)^{\text{H}} \mathbf{H}_{m,n}\right)\right), \\
&= -\operatorname{tr}\left(\left(\mathbf{H}_{m,n}^{\text{meas}}\right)^{\text{H}} \mathbf{H}_{m,n}^{\text{meas}}\right) - \frac{3}{2} \cdot \frac{a_{m,n}^2}{d_{m,n}^6} + \frac{2a_{m,n}}{d_{m,n}^3} \operatorname{tr}\Big(\underbrace{\text{Im}\left\{\mathbf{H}_{m,n}^{\text{meas}}\right\}^{\text{T}} \mathbf{O}_n^{\text{T}}}_{\mathbf{A}_{m,n}} \underbrace{\mathbf{F}_{m,n} \mathbf{O}_m}_{\mathbf{B}_{m,n}}\Big),
\end{aligned}$$

where we used (6.2), the cyclic property of the trace, and $\operatorname{tr}(\mathbf{F}_{m,n}^{\text{H}} \mathbf{F}_{m,n}) = \frac{3}{2}$. Since the first summand is independent of $\boldsymbol{\psi}_m$, the maximization of the log-LHF thus simplifies to

$$(\hat{\boldsymbol{\psi}}_m^{\text{ML}})_{3\text{ax}} = \arg\max_{\boldsymbol{\psi}_m} -\frac{3a_{m,n}^2}{2d_{m,n}^6} + \frac{2a_{m,n}}{d_{m,n}^3} \operatorname{tr}\left(\mathbf{A}_{m,n} \mathbf{B}_{m,n}\right), \tag{6.25}$$

where $\mathbf{A}_{m,n} \overset{\text{SVD}}{=} \mathbf{U}_{m,n} \mathbf{S}_{m,n} \mathbf{V}_{m,n}^{\text{T}}$ is fixed based on the measured channel matrix $\mathbf{H}_{m,n}^{\text{meas}}$ and the known anchor orientation matrix $\mathbf{O}_n$. Moreover, $\mathbf{B}_{m,n}$ is the optimization variable under the structural constraints of both $\mathbf{F}_{m,n}$ and $\mathbf{O}_m$.

We consequently need to maximize $\operatorname{tr}(\mathbf{A}_{m,n} \mathbf{B}_{m,n})$ independently of the distance. According to Von Neumann's trace inequality [133], this term is generally maximized if both matrices share the same singular vectors. However, this may not be feasible when $\mathbf{O}_m$ is required to be a proper rotation matrix. Our problem hence becomes a specific constrained form of the Procrustes problem (cf. [134]) and is closely related to the Kabsch algorithm [135]. It's solution is obtained via (cf. Appendix B)

$$\hat{\mathbf{O}}_m^{\text{ML}} = \mathbf{V}_{m,n} \operatorname{diag}\left(1, -1, -\det\left(\mathbf{U}_{m,n} \mathbf{V}_{m,n}^{\text{T}}\right)\right) \mathbf{U}_{m,n}^{\text{T}}, \tag{6.26}$$

$$\hat{\mathbf{F}}_{m,n}^{\text{ML}} = \mathbf{V}_{m,n} \operatorname{diag}\left(1, -\frac{1}{2}, -\frac{1}{2}\right) \mathbf{V}_{m,n}^{\text{T}} \tag{6.27}$$

and hence the maximum value of the trace is equal to $z_{m,n} = \operatorname{tr}\left(\mathbf{S}_{m,n} \operatorname{diag}\left(1, \frac{1}{2}, \frac{1}{2} \cdot \det(\mathbf{U}_{m,n} \mathbf{V}_{m,n}^{\text{T}})\right)\right)$. For any given $d_{m,n}$, this solution

[1] We limit ourselves to the agent-anchor case to simplify the maximization of (6.25). Alternatively, a straightforward extension to the agent-agent case would require to also consider $\mathbf{O}_n$ in the joint maximization and may lead to ambiguities of the solution for the estimates of the agent orientation.

maximizes the LHF while satisfying the structural constraints of both estimated matrices. The corresponding ML distance estimator is then found by solving $-\frac{9a_{m,n}^2}{d_{m,n}^7} + \frac{6a_{m,n}z_{m,n}}{d_{m,n}^4} \overset{!}{=} 0$, which yields

$$(\hat{d}_{m,n}^{\mathrm{ML}})_{3\mathrm{ax}} = \sqrt[3]{\frac{3}{2} \cdot \frac{a_{m,n}}{z_{m,n}}}, \tag{6.28}$$

where the real-valued root is chosen as estimate.

In addition, the previous solutions entail another interesting result, since $\hat{\mathbf{F}}_{m,n}^{\mathrm{ML}}$ offers a direct estimate of the unit direction vector between agent and anchor. That is, up to an ambiguity of its sign, the singular vector of $\mathbf{V}_{m,n}$ which corresponds to the largest singular value can be chosen as $\hat{\mathbf{u}}_{m,n}^{\mathrm{ML}}$, i.e. $\hat{\mathbf{u}}_{m,n}^{\mathrm{ML}} = \mathbf{V}_{m,n}[1,0,0]^{\mathrm{T}}$. As a result, we also obtain an estimate on the agent position as

$$(\hat{\mathbf{p}}_m)_{3\mathrm{ax}} = \mathbf{p}_n \pm \hat{\mathbf{u}}_{m,n}^{\mathrm{ML}} (\hat{d}_{m,n}^{\mathrm{ML}})_{3\mathrm{ax}} \,. \tag{6.29}$$

In practice, we empirically choose to resolve the ambiguity by deciding on the solution that is within the room and otherwise discarding it. In combination with the orientation estimate $\hat{\mathbf{O}}_m^{\mathrm{ML}}$, which can be transformed to an estimate of $[\alpha_m, \beta_m, \gamma_m]_{3\mathrm{ax}}^{\mathrm{T}}$ by the corresponding trigonometric relationships, we can ultimately estimate the full agent deployment $(\boldsymbol{\psi}_m)_{3\mathrm{ax}}$ based on each single measurement of the channel matrix between an agent and an anchor. Compared to the solenoid case, we thus do not even require an additional multilateration and hence no numerical solvers to obtain a position estimate. However, if the observations of multiple anchors are to be combined, it is clearly possible to use the corresponding distances estimates of (6.28) as input for an analogous multilateration as in (6.24) for three-axis anchors, where the weights are determined on the basis of $\|\mathbf{H}_{m,n}\|^2$ instead of $|h_{m,n}|^2$.

6.3.3 Advanced LS Initialization

The previously introduced pairwise estimation approaches can either be used as position estimators in their own right or the resulting estimates may serve as a starting point for the more complex joint LS estimation. Having an improved starting point close to the global minimum of the highly-dimensional cost function can reduce the required processing time as well as improve the chances of even finding the global minimum at all. We will hence also consider an estimation approach where both, the estimates of Sec. 6.3.2 are used as initialization values for the cooperative and non-cooperative LS estimators of Sec. 6.3.1. For three-axis coils we initialize with the obtained position and orientation estimates, and only use the single estimate of the closest anchor, i.e. the one which yields the smallest distance estimate. Should this anchor not have a unique valid position estimate, i.e. both position estimates are either outside or within the room due to the ambiguity of (6.29), then we iteratively choose the anchor which has the next smallest distance estimate. For solenoid coils only the position estimates are available and the orientation variables are hence initialized as uniformly random points on the unit sphere.

6.4 Performance Evaluation for IoT Topologies

We next look at an IoT motivated setup which comprises a 1.5 m cubic volume with fixed *anchors* on the boundaries (e.g. walls) and multiple randomly placed *agents* within the volume. In Fig. 6.1a we have illustrated the setup with $M = 7$ solenoid agents and $N = 12$ solenoid anchors, whereas Fig. 6.1b shows the same cubic room with $M = 7$ three-axis agents and $N = 4$ three-axis anchors. Throughout Sec. 6.4, the anchors remain exactly as depicted in Fig. 6.1 while a variable number M of randomly deployed agents is considered. More specifically, the agents' orientations are chosen to be i.i.d. uniform random points on the unit sphere and the agents' placements are uniformly random within the room, but restricted such that they satisfy a minimum distance constraint ($d^{\text{min}} \geq 3D^{\text{coil}}$) between any coil pair. The minimum distance constraint is higher than the ones used in Cpt. 4 and Cpt. 5 as it is not only used to prevent strong coupling and the proximity problem, but also to justify the unilateral assumption and ideal power matching.

The large cubic setup leads to longer agent-anchor distances on average, compared to the setups considered in Cpt. 4 and Cpt. 5. However, depending on the IoT devices used it is now also possible to employ larger coil antenna diameters to mitigate the associated increase of the path loss. For the numerical evaluations of this chapter, all identical coils (and subcoils) hence have an increased diameter of $D^{\text{coil}} = 5\,\text{cm}$. Yet, we only use $N^{\text{coil}} = 5$ coil windings to still allow for a flat coil design that may conveniently fit inside an IoT device. Additionally, the operating frequency is reduced to $f =$

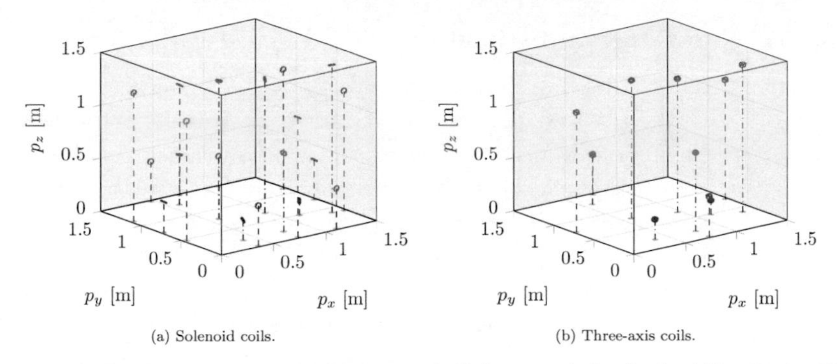

(a) Solenoid coils. (b) Three-axis coils.

Figure 6.1: 3D cubic room setup with $M = 7$ agents (red) that are to be localized and $N = 12$ solenoid anchors (blue, left) or $N = 4$ three-axis anchors (blue, right).

coil turns	N^{coil}	5
coil diameter	D^{coil}	50 mm
coil height	H^{coil}	10 mm
wire diameter	D^{wire}	1 mm
design frequency	f^{des}	500 kHz
op. frequency	f	500 kHz
conductivity	σ^{wire}	59.6MS/m
rel. permittivity	ϵ_{r}	1
rel. permeability	μ_{r}	1
noise variance	σ^2_{sol}	$\frac{1}{3} \cdot 10^{-10}$
noise variance	σ^2_{3ax}	$1 \cdot 10^{-10}$

(a) Specified parameters.

resistance	R	0.07 Ω
self-inductance	L	1.7 µH
self-capacitance	C	2 pF
wire length	l^{wire}	0.8 cm
wavelength	λ	600 m
coil Q-factor at f^{des}	Q	75
coil self-resonance	f^{self}	86 MHz

(b) Resulting parameters.

Table 6.1: Simulation parameters and resulting quantities.

500 kHz to further improve material penetration. The resulting system parameters[2] and other specifications are listed in Tab. 6.1. For all numerical optimizations we employ the Levenberg-Marquardt algorithm (cf. [136]) via Matlab's built-in *lsqnonlin* function.

[2]Note that if using the same parameters as in Cpt. 4 and Cpt. 5 (cf. Tab. 4.1) while accounting for our choice $x_{m,n} = 1$ from (6.3), we would obtain $\sigma^2_{\text{sol}} \approx 2.5 \times 10^{-12}$ for the noise variance. However, this quantity is drastically affected by the quality of the measurement device and the measurement duration, which may vary depending on the field of application. To make our results comparable to other existing literature in the IoT field, we instead choose $\sigma^2_{\text{sol}} = \frac{1}{3} \times 10^{-10}$. This choice is in line with [61], who found this noise level to be appropriate for office environments based on vector network analyzer measurements with coils similar to ours.

6.4.1 Numerical LS Estimation

In Fig. 6.2, it is shown how the accuracy for both cooperative (black) and non-cooperative (gray) localization schemes changes with an increasing number of agents. For each number of agents M, 1000 random agent topologies were considered and the PEB and RMSE were calculated for each one. The triangle-lines represent the corresponding mean PEBs of agent 1 for each M, whereas the cross-lines illustrate the mean RMSE of the corresponding LS estimators, where the mean is taken over the random agent topologies. The numerical LS estimators (numLS) used the true agent deployment as starting points. For both coil types, we see that the minimum which is found on average with this perfect initialization corresponds to the PEB. Without cooperation between the agents, the performance is not affected by their number M and remains constant. For the cooperative scenario we see a steady improvement in both scenarios that result in a comparable relative improvement. We will refer to this improvement as cooperation gain. This cooperation gain can be attributed to three factors: (i) to having more channel observations, similarly to the noise-averaging effect when having more anchors, (ii) to having stronger channel gains between agent pairs that are in close proximity, which for solenoid coils may also counterbalance agent-anchor misalignment losses, and (iii) to obtaining more directional information on the agent position. In Fig. 6.3 we emphasize (ii) by the showing the CDFs of the squared magnitudes of the channel gains $[\mathbf{H}_{m,n}]_{k,l}$ between any pair of either agent-agent or agent-anchor three-axis subcoils for $M = 10$ randomly placed agents within the room. We observe that about 4% of the channel gains are below the corresponding noise level and that the agent-agent channel gains are generally stronger than the agent-anchor channel gains. In practice, the stronger channel gains are especially relevant for the solenoid setup, as the additional cooperation helps to mitigate the orientation-based alignment loss similarly to passive relays, i.e. we can obtain position information about an agent via its surrounding peers even if this agent itself is misaligned to all anchors.

Thanks to the initialization with the true deployment vector, the numerical solvers in Fig. 6.2 approach the PEB. Without an unrealistic perfect initialization, it remains unclear however, whether the cooperation gain found is actually feasible. To investigate this matter, Fig. 6.4 shows the corresponding CDFs of the Euclidean norms of all instantaneous positions errors for the case of $M = 10$ randomly deployed agent nodes (multiple agent topologies and noise realizations). In Fig. 6.4a - Fig. 6.4d, we distinguish between the perfect initialization (*perf. init.*) and a random initialization

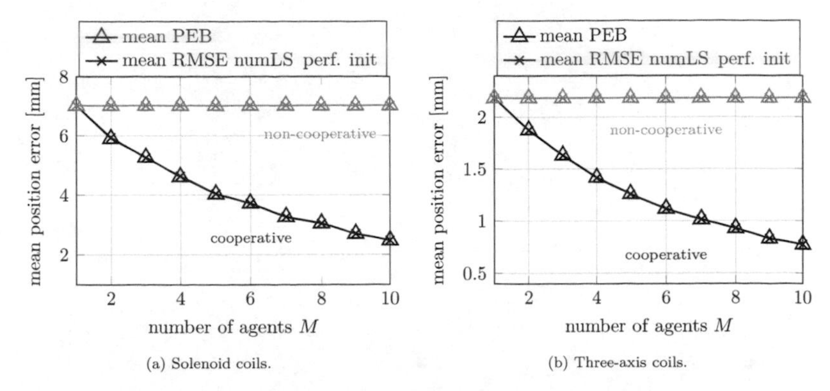

(a) Solenoid coils.

(b) Three-axis coils.

Figure 6.2: Comparison of the localization accuracy for cooperative vs. non-cooperative approaches with an increasing number of agents in the room. The numerical ML estimators are perfectly initialized and their resulting root-mean-square errors tightly approach the corresponding PEBs.

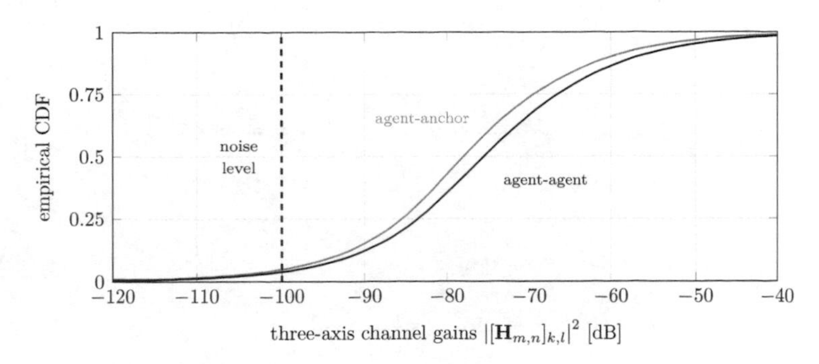

Figure 6.3: CDFs of the magnitude squared channel gains for $M = 10$ randomly deployed three-axis agents in the setup.

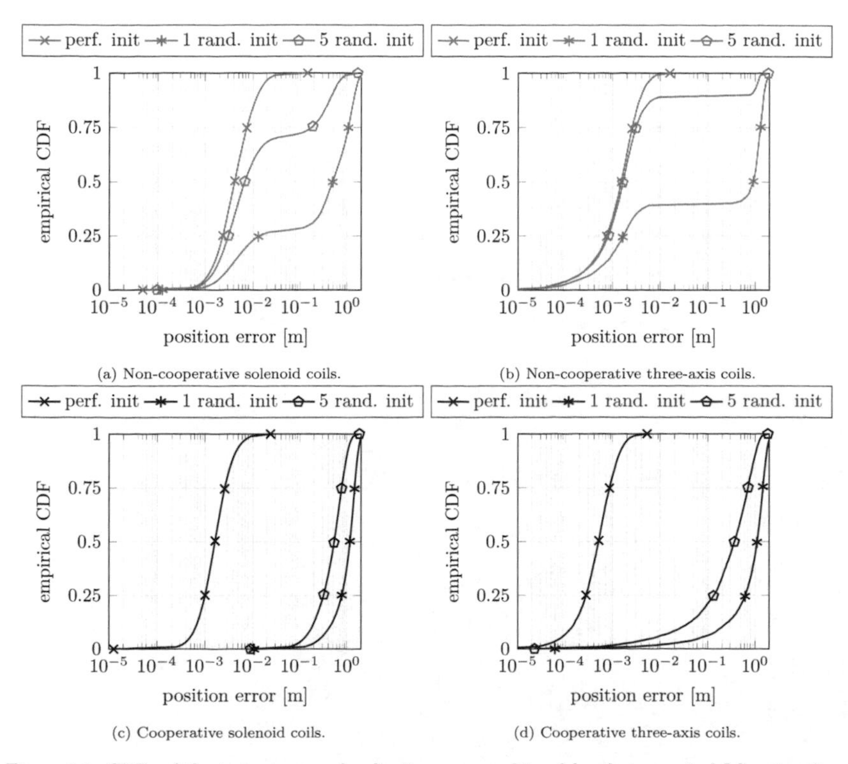

Figure 6.4: CDFs of the instantaneous localization errors achieved by the numerical LS estimation (numLS) for different initializations.

(*rand. init.*) within the room with either 1 or 5 random initializations[3]. For the easier non-cooperative localization in Fig. 6.4a, we observe that a single random initialization only suffices in about 27% of the cases for solenoids, 40% for three-axis coils and otherwise converges to local minima. With 5 random initialization, the percentage of outliers reduces to roughly 25% or 10%, respectively. With even more random initialization the outliers could possibly be rendered insignificant. For the specific case of non-cooperative solenoids, [26] showed that alternating optimization approaches, which switch between position and orientation optimization, achieve a better convergence behavior than trying to optimize all parameters simultaneously via gradient-based search. For the cooperative localization in Fig. 6.4c and Fig. 6.4d, we see that additional random initializations also generally improve the performance, but even with 5 random initializations we still do not even remotely approach the global cost minimum of the perfectly initialized case for either coil type. Even worse, despite combining cooperation and multiple random initializations, the median performance is worse than the one of the corresponding non-cooperative approach. Moreover, using five random initializations also entails a five-fold complexity increase, which is especially problematic for the highly-dimensional cooperative joint optimization. This shows that the increased complexity of the optimization problem can directly nullify the cooperation gain which the additional agent-agent observations otherwise provide.

6.4.2 Reduced Complexity Estimators

Next, we thus compare the performance of the reduced-complexity estimators, which have been proposed in Sec. 6.3.2 and Sec. 6.3.3. The estimators are also summarized in Tab. 6.2 and a flow chart representation for each approach is provided in Tab. 6.2. The color of the flow chart represents the coil affiliation, i.e. blue elements are suitable for or belong to solenoid coils, red elements only for three-axis coils and purple elements correspond to either coil type independently. We study their performance for the same cubic setup of Fig. 6.1 with $M = 10$ possibly cooperating agents.

Position Error In Fig. 6.6, we show the empirical CDFs of the position errors obtained with each of these approaches. For solenoid case in Fig. 6.6a, we see that the turboLS schemes (squares) do not yield the same results as the perfectly initialized ones (crosses), i.e. applying a prior multilateration does not fully solve the

[3]The ultimately selected position estimate was the one that corresponded to the lowest cost over all individual optimization processes

Scheme	Description
numLS	A numerical solver (Levenberg-Marquardt algorithm) with a certain initial parameter value is applied to the cost functions from (6.19) to (6.22) in an attempt to solve the associated LS problem.
multilateration	Obtain ML distance estimates between an agent and all anchors with the closed-form rule (6.23) for solenoid coils and (6.28) for three-axis coils. Then estimate the agent position via multilateration, i.e. (6.24) or analogously for three-axis coils (implemented via gradient search).
pairML	Position and orientation estimates of an agent are calculated via the closed-form ML estimation rules (6.26) to (6.29), based on the measurement of the anchor with the smallest ML estimated distance. This approach is only viable for three-axis coils.
turboLS	For the three-axis coils, the position and orientation estimates obtained with pairML are used as initial parameter values in numLS, which functions as afterburner. For the solenoid coils, the results of the multilateration are used as initial parameters.

Table 6.2: Summary of all proposed localization schemes.

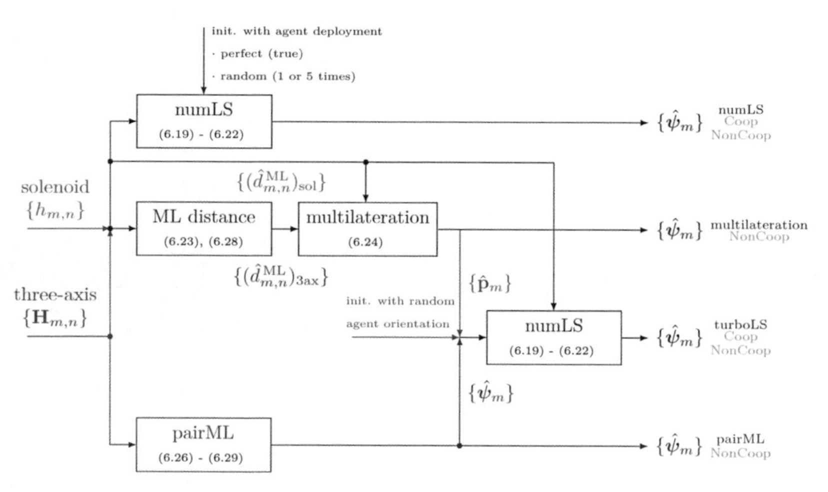

Figure 6.5: Conceptunal flow chart of all proposed localization schemes.

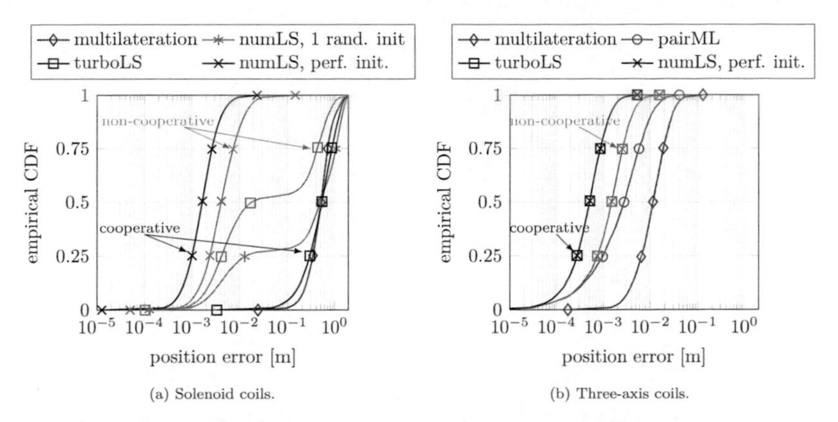

(a) Solenoid coils.

(b) Three-axis coils.

Figure 6.6: CDF comparison of the instantaneous localization errors achieved for all proposed localization estimators.

convergence issues of the numerical optimization process. Nevertheless, at least for the non-cooperative case the performance improves compared to the numLS approach with one random initialization (asterisk). For the cooperative case, the turboLS however yields a comparable performance as the multilateration (diamonds) with which it is initialized, so the numerical afterburner has no positive impact.

For the three-axis setup in Fig. 6.6b, we see that for both the cooperative as well as the non-cooperative scenarios, the turboLS schemes achieve the same performance as the perfectly initialized numLS ones. Without the LS afterburner, the pairML scheme (circles) maintains an acceptable performance that is comparable to that of the non-cooperative approaches despite its ultra-low complexity and despite only requiring a single agent-anchor link. Moreover, it yields a median position error of about 3 mm, which may be sufficient for various practical applications. Lastly, we find that multilateration (diamonds) exhibits poor accuracy compared to all other approaches.

Computation Time In order to further emphasize on the differences in computational complexity, Fig. 6.7 shows the CDFs of the time that each approach requires in order to fully localize $M = 10$ agents. To this end, the optimization processes were run on an Intel Core i7-10750H 2.6 GHz processor in the single core mode. For the solenoid setup (Fig. 6.7a), the multilateration requires less than 40 ms during most localizations and the other non-cooperative and cooperative approaches require 2.6 s or 18 s, respectively. As there is no reduction of the computation time when using either the numLS or the turboLS approach for solenoid coils, the multilateration is a clear favorite for these coil types when there is no cooperation and a low-complexity requirement. For the three-axis setup (Fig. 6.7b), the pairML approach requires less than 18 ms in almost all cases and the multilateration less than 25 ms. For the LS approaches, we observe that the non-cooperative turboLS mostly has a processing time of less than 153 ms and its cooperative equivalent requires roughly 13 s. We also see that these turboLS approaches, which are required to reliably find the global minimum, reduce the median processing time compared to the numLS solution even with a single random initialization. Nevertheless, the required processing time of any cooperative approach compared to its non-cooperative alternative is still increased by more than an order of magnitude.

For a concise overview, Fig. 6.8 further links the results of the computation time to the corresponding position errors of each approach by showing the median of each quantity via scatter plot, both for solenoid coils and three-axis coils. Additionally, the dashed lines represent the cooperative or non-cooperative median PEBs as theoretical upper limit on the performance. For solenoid coils, Fig. 6.8a clearly visualizes that the cooperation gain cannot be realized by the considered estimators and the practical choice between estimators is hence limited to the multilateration (low-complexity) and the non-cooperative turboLS approach (high accuracy). For three-axis coils we find via Fig. 6.8b that the same trade-off between complexity and accuracy holds but in this case the cooperation gain can be realized and there are overall more viable choices available. In detail, the lowest complexity is offered by the pairML approach, which also shows a higher accuracy than either multilateration or numLS with one random initialization. If a higher accuracy is required it is possible to switch to the non-cooperative turboLS approach, which halves the median position error in exchange for an order of magnitude more computation time. Lastly, if complexity is not an issue the cooperative turboLS approach can be chosen. This choice decreases the median position error by a factor of 4 compared to the pairML approach but requires a computation time that is more than two orders of magnitude higher, on top of also needing a cooperating network.

6.5 Conclusions

In this chapter, we studied the impact of agent cooperation on MI localization. That is, we stated the PEB if the cooperation of agents allows us to obtain noisy inter-agent channel gains when using either solenoid or three-axis coils. For an IoT motivated

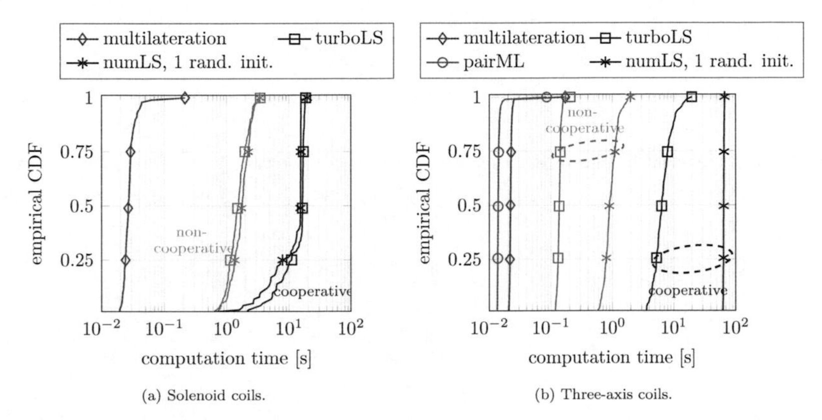

(a) Solenoid coils. (b) Three-axis coils.

Figure 6.7: CDF comparison of the time which each approach requires to fully localize a network comprising $M = 10$ agents.

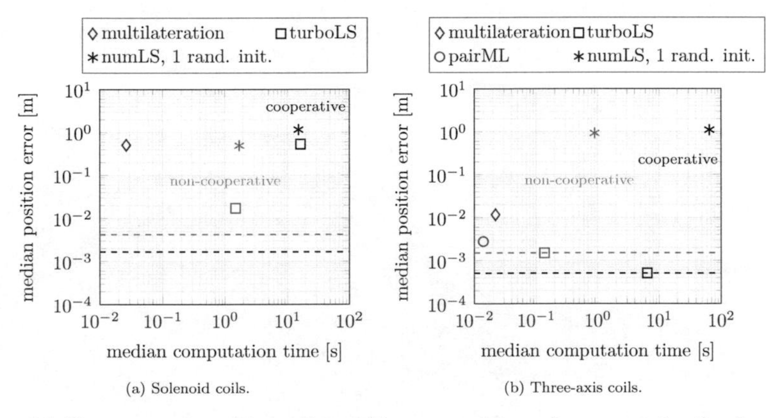

(a) Solenoid coils. (b) Three-axis coils.

Figure 6.8: Direct comparison of the median position error and the median computation time for each approach when localization a network comprising $M = 10$ agents.

setup, we then examined the theoretical cooperation gain in dependence of the number of agents. However, common numerical estimators failed to put this theoretical cooperation gain into practice, as cooperation increased the complexity of the associated optimization problem and caused convergence issues. In a first step to circumvent this issue, we hence derived the closed-form ML deployment estimator for a single pair of three-axis coils. In a second step, we combined this closed-form solution with a numerical afterburner which allowed us to bypass the convergence issues an tightly approach the theoretical gain. Yet, the required computation time for this approach far exceeds that of its non-cooperative counterpart. Cooperative MI localization may thus only be viable for applications that provide abundant computation power or are not time-sensitive.

Chapter 7

Topology Classification Using Purely Passive Tags

Note: Parts of this chapter have been published by us in [91]. This work hence exhibits similarities regarding formulations and visualizations.

In this chapter, we will propose and investigate a novel low-complexity idea to indirectly determine a network constellation (also network topology) $\boldsymbol{\Psi}$ of a dense MI network. We will first describe this idea conceptually, then elaborate on important design choices, provide possible practical use cases, and lastly adapt the full system model from Sec. 3.1 to better match the new concept. In Cpt. 8 and Cpt. 9 this idea is further investigated for posture recognition.

7.1 Topology Classification System Concept

The proposed system uses resonantly loaded and purely passive agents (identical to purely passive relays), whose deployment vectors we do not need to know or estimate. To make this switch and distinction clear, we will call them *passive tags*. Additionally, we will also deploy anchors but in contrast to Cpt. 3 to Cpt. 6 do not need to know their deployment vectors either.

We place a multitude of such passive tags and anchors on an object or in an area of interest. This placement may either be performed randomly or systematically. Next, when we measure the input impedance of an anchor, we automatically feed a current (or voltage) with a constant amplitude into it, which generates a magnetic field. This magnetic field further induces currents in all nearby passive tags, which in turn generate new magnetic fields that superimpose the old one. The presence of the passive tags hence changes the observable impedances of all anchors, as was already evident via (3.8). Moreover, the entire set of passive tags can alternatively also be interpreted as a single combined tag that changes its form and coupling characteristics with each different topology of its distributed constituent parts. As a result of this dependency on the relative positions and alignments between all coils, each different network constellation leads to a possibly unique set of detuned input impedances that can be used as a fingerprint. Learning these fingerprints via supervised classifiers during an initial calibration phase allows for later re-associations and therefore enables the ability to classify between different network topologies. The novelty of this idea lies in the use of purely passive tags that do neither require batteries nor ICs, and whose deployment is unknown. Moreover, for simple networks with few topologies that need to be distinguished, the system may already work with a single anchor and single-frequency impedance measurements. Still, the extent of the detuning is limited by the coil designs and their placements, which in turn are limited by practical constraints. It is thus uncertain whether the passive impedance changes are even noticeable under the presence of additional noise for any given use case.

7.2 General Design Challenges and Considerations

While this general idea may be useful for a wide variety of applications, some important limitations need to be considered.

1. Passive tags can only noticeably affect and detune anchors to which they are strongly coupled. For the coils used in Cpt. 4 to Cpt. 5, this required a separation of less than 3 coil diameters (cf. Fig. 4.1). For an increased coverage of the system, it is hence beneficial to (i) distribute the anchors such that there are always passive tags in close vicinity (cf. Fig. 4.1c), (ii) use coils that have high quality factors (cf. Fig. 4.1d), (iii) exploit dense swarms of passive tags such that the compound interaction between them boosts the coupling range to the anchors similar to a waveguide effect (cf. Fig. 5.5), and (iv) use three-axis coils as in Cpt. 6 to prevent misalignments.

2. All moving parts or objects that are to be tracked need to have at least one passive tag or anchor attached to them, such that their movement elicits a detuning. Higher numbers of passive tags may however be beneficial as their inter-tag coupling leads to more intricate and possibly unique detuning patterns.

3. The directional information of a passive tag still affects the system, i.e. tangential movements of a tags typically have a weaker impact on the impedance than radial movements (cf. Fig. 4.5). If possible, it is hence beneficial to use distributed instead of clustered anchors setups. Alternatively, using swarms of passive tags that are well-distributed also helps to enhance the impact of tangential movements.

4. The impact of the passive tags may be weak compared to that of an active interferer, so the choice of a fitting operating frequency and bandwidth is crucial. Alternatively, interference cancellation methods have to be applied. Additionally, the system should be operated in low-distortion environments since nearby ferrous materials have an irregular impact on the detuning that cannot always be calibrated for.

7.3 Possible Use Cases for Topology Classification

In the following, we offer a brief list of examples where the proposed topology classification system with spreaded passive tags may be applied.

Body Posture and Activity Recognition Body posture and activity recognition are eHealth applications that are envisioned to prevent the emergence of MSDs by recognizing unhealthy postures, to monitor rehabilitation progress after injury, to contact emergency services in case of falls, or to provide detailed fitness tracking which may improve athletic performance.

The proposed system may operate as a low-cost and low-complexity posture recognition system that omits the need of battery powered sensors placed all over the body. That is, multiple anchor coils may for example be centralized on the torso with passive tags being situated on all limbs. This setup would limit the measurement process to the torso while still enabling full body posture recognition. The coils can furthermore be integrated into clothing for an improved ease of use. This use case is further studied in Cpt. 8 and Cpt. 9.

Adhesive Stripes Various passive tags may be integrated into any kind of adhesive stripes (e.g. sticky tape, duct tape, or adhesive bandages) to monitor changes of their form or to determine relative changes of the positions and orientations between multiple tag-integrated stripes. These stripes may for example be applied to single body parts such as the neck or ankles to alert users in case of specific malalignments. Alternatively, stripes such as duct tape may be used for repairs within obstructed areas, such that their status can afterwards be monitored from the outside by means of an external anchor.

Surgical Meshes Similar to the idea of adhesive stripes, surgical meshes or other fixed implants may be equipped with passive tags before placing them within the body.

An external sensor head comprising anchors may then be used for regular check ups to examine whether the location and form of the inserted meshes or implants are still correct.

7.4 Simplified System Model

We next adapt the full system model from Sec. 3.1 to better fit the idea from Sec. 7.1. First, the adjusted model entirely leaves out the switchable passive relays and is only composed of anchors and passive tags, i.e. strictly passive agents that are loaded with a fixed capacitance to make them resonant in the absence of other coils. In contrast to the general passive relays, the resonantly loaded passive tags do not contain a switch and hence cannot be switched off. Second, the anchors now only operate iteratively (time multiplexed) by using switches and are connected to individual lossless two-port matching networks that each power match the corresponding coil antenna impedance $Z_{\mathrm{R},1},\ldots Z_{\mathrm{R},N_{\mathrm{R}}}$ to a reference value R^{ref} for an impedance measurement device. Lastly, in Sec. 3.1 we used the anchors' input current vector $\mathbf{i}_{\mathrm{R}}^{\mathrm{in}}$ as observation quantity, since it allowed for consistency between the active and passive agent case. However, this consistency is not required anymore as we only consider passive agents. We therefore opt to omit the otherwise required current division (cf. (3.18) and (3.20)) and switch to measurements of the individual anchor input impedances $Z_{\tilde{\mathrm{R}},1}^{\mathrm{in}},\ldots,Z_{\tilde{\mathrm{R}},\mathrm{N_R}}^{\mathrm{in}}$. Practically, these impedances may easily be measured by means of an impedance meter or impedance converter systems [137, 138]. A circuit model of this simplified system can be found in Fig. 7.1.

Since we measure the anchor's impedances sequentially (i.e. only a single switch in Fig. 7.1 is closed at a time), the scalar coupling input impedance $Z_{\mathrm{C,n_R}}^{\mathrm{in}}$ at port n_{R} is found as (cf. (3.8))

$$Z_{\mathrm{C,n_R}}^{\mathrm{in}} = [\mathbf{Z}_{\mathrm{C}:\tilde{\mathrm{R}}}]_{n_{\mathrm{R}},n_{\mathrm{R}}} - [\mathbf{Z}_{\mathrm{C}:\tilde{\mathrm{R}}\tilde{\mathrm{T}}}]_{n_{\mathrm{R}},:}\,(\mathbf{Z}_{\mathrm{C}:\tilde{\mathrm{T}}} + \mathbf{Z}_{\mathrm{T}}^{\mathrm{load}})^{-1}\,[\mathbf{Z}_{\mathrm{C}:\tilde{\mathrm{R}}\tilde{\mathrm{T}}}]_{n_{\mathrm{R}},:}^{\mathrm{T}}\,, \tag{7.1}$$

In case the self capacitances of all coils are negligible, the scalar $[\mathbf{Z}_{\mathrm{C}:\tilde{\mathrm{R}}}]_{n_{\mathrm{R}},n_{\mathrm{R}}}$ coincides with the coil impedance of the corresponding anchor and the vector $[\mathbf{Z}_{\mathrm{C}:\tilde{\mathrm{R}}\tilde{\mathrm{T}}}]_{n_{\mathrm{R}},:}$ simply comprises the mutual impedances between this anchor and all passive tags. Otherwise the remaining open-circuited anchors need to be incorporated into the matrix $\mathbf{Z}_{\mathrm{C}}$ analogously to open-circuited passive relays (cf. (5.2)).

The adjacent lossless matching networks are considered to be T-structured two-

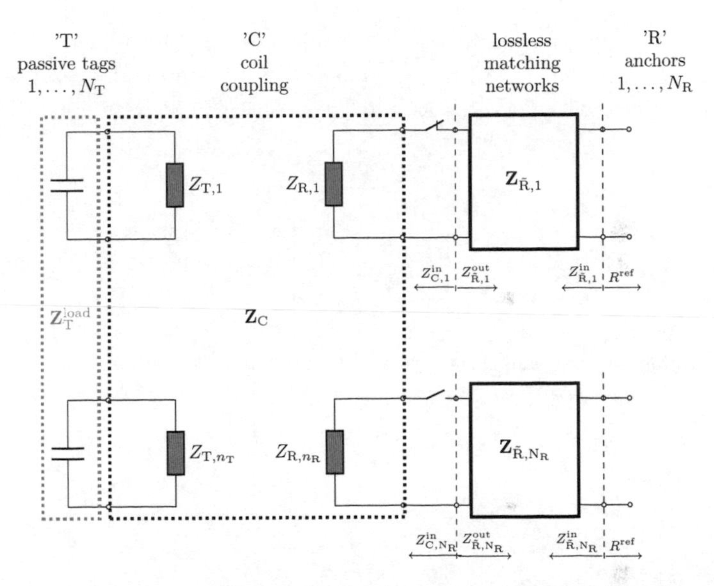

Figure 7.1: Multiport circuit representation of passive tags and anchors with their their impedance matrix and individual lossless matching networks.

port power matching networks that match the measurement devices resistance[1] R^{ref} to the coil antenna's impedance $Z_{\mathrm{R},n_\mathrm{R}}$ (cf. (2.10)). They thus ultimately transform the coupling input impedance of $Z^{\mathrm{in}}_{\mathrm{C,n_R}}$ to the anchor input impedance $Z^{\mathrm{in}}_{\tilde{\mathrm{R}},\mathrm{n_R}}$ via (cf. (3.9) and (3.31))

$$Z^{\mathrm{in}}_{\tilde{\mathrm{R}},\mathrm{n_R}} = R^{\mathrm{ref}}\mathrm{Re}(Z_{\mathrm{R},n_\mathrm{R}})\left(Z^{\mathrm{in}}_{\mathrm{C,n_R}} - j\mathrm{Im}(Z_{\mathrm{R},n_\mathrm{R}})\right)^{-1}. \tag{7.2}$$

If no other coils are within range of an anchor n_R, it is clear that $Z^{\mathrm{in}}_{\mathrm{C,n_R}} = Z_{\mathrm{R},n_\mathrm{R}}$ and hence $Z^{\mathrm{in}}_{\tilde{\mathrm{R}},\mathrm{n_R}} = R^{\mathrm{ref}}$, as intended.

By combining the individual anchor observations, we can obtain the possibly detuned impedance vector $\mathbf{x}_p^0$ and the corresponding feature vector $\mathbf{r}_p^0$ for a fixed deployment of all coils, i.e. for a specific network constellation $\mathbf{\Psi}_p$ with $p = 1, \ldots, N_p$, as

$$\mathbf{x}_p^0 = \begin{bmatrix} Z^{\mathrm{in}}_{\tilde{\mathrm{R}},1} \\ Z^{\mathrm{in}}_{\tilde{\mathrm{R}},2} \\ \vdots \\ Z^{\mathrm{in}}_{\tilde{\mathrm{R}},\mathrm{N_R}} \end{bmatrix}, \quad \mathbf{r}_p^0 = \mathrm{vec}\left(\begin{bmatrix} \mathrm{Re}(\mathbf{x}_p^0)^\mathrm{T} \\ \mathrm{Im}(\mathbf{x}_p^0)^\mathrm{T} \end{bmatrix}\right). \tag{7.3}$$

The separation into real and imaginary parts is used in order to simplify the implementation of classification algorithms.

[1]In Cpt. 8 we choose $R^{\mathrm{ref}} = 50\,\Omega$ to be in line with our **V**ector **N**etwork **A**nalyzer (VNA) measurements in Cpt. 9. However, this value may change significantly depending on the specific application and measurement device used.

Chapter 8

Human Posture Recognition: A Case Study

Note: Parts of this chapter have been published by us in [91]. This work hence exhibits similarities regarding formulations and visualizations.

In this chapter, we examine the low-complexity topology classification system of Cpt. 7 for human body posture recognition. The main goal is to assess the system's general feasibility for this use case via simulation study. We further want to determine the relevance of basic parameters such as the number of required coils and their design.

To this end, we first introduce human body models in Sec. 8.1 which consist of rigid, three-dimensional body parts that are interconnected by rotatable joints. These models allow us to dynamically place and move different on-body coils and matching network, such that realistic postures and hence spatial coil configuration can be simulated. We then introduce four different types of perturbations to our coupling model from Sec. 7.4 in order to efficiently study the system's noise sensitivity. After combining the MI system and the body model, we further compare two different types of coil designs and investigate their practical differences for 14 everyday postures. Subsequently, the generated noisy datasets are used in Sec. 8.5 to train and evaluate well-known supervised classification algorithms.

8.1 Body Models and Postures

With the MI model being defined in Sec. 7.4, the next step is to introduce human body models on which the anchors and passive tags can be placed. We further choose 14 different everyday postures, which these body models have to re-enact and which ultimately need to be distinguished by the posture recognition system.

8.1.1 Joint-Based Human Body Model

The considered human body models (sometimes referred to as *mannequins*) consist of ten joints that are connecting different parts of the body. More precisely, the model includes the body parts: calves, thighs, pelvis, torso, head, arms, forearms. Each of

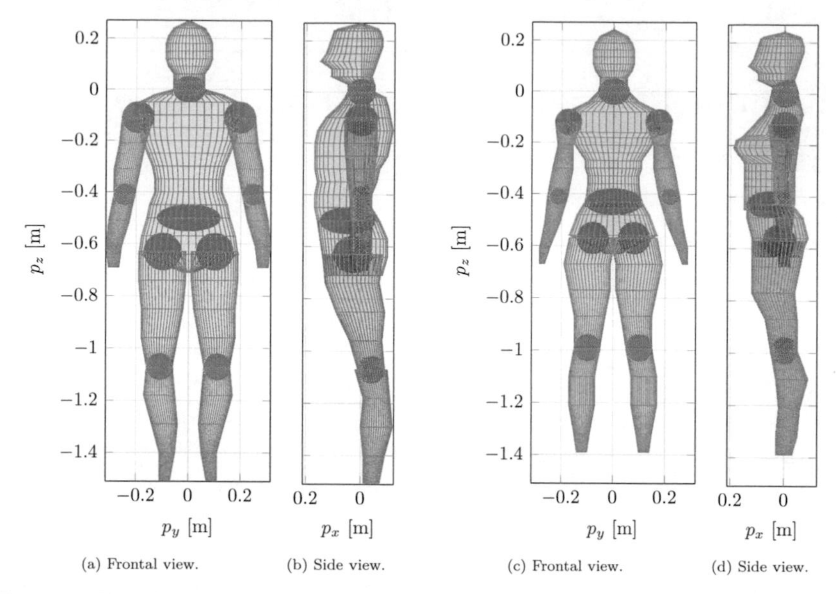

(a) Frontal view. (b) Side view. (c) Frontal view. (d) Side view.

Figure 8.1: Frontal and side views of the human body models Martin and Fiona, each consisting of 10 rotatable joints (green) and 11 rigid body parts (gray mesh polygons).

these body parts is connected to at least one other body part by one of the ten joint angles as illustrated in Fig. 8.1. In Fig. 8.1a and Fig. 8.1b we show the fit male body model called *Martin* (M), whereas Fig. 8.1c and Fig. 8.1d show the analogous fit female body model called *Fiona* (F).

Due to the rigid nature of the body parts, the distance between any two directly connected joints remains constant, independent of the posture. The orientation of each joint is defined with individual roll, pitch, and yaw angles, except for the hinge joints (elbow and knee joints), which are each defined by only one angle. For any given posture, the position and orientation of any body part are hence defined solely by the orientations (angles) of all joints and can be calculated easily as a concatenation of coordinate transformations, beginning at the torso.

8.1.2 Postures

All the postures $p \in \{1, \ldots, N_\mathrm{p}\}$ with $N_\mathrm{p} = 14$ which are investigated in this work are briefly summarized in Tab. 8.1 with a simplified sketch being presented in Fig. 8.2. These include five sitting and nine standing postures. Moreover, three of these postures

	Posture	Description
Sit	I	Sitting with the arms hanging down
	Front	Sitting with the arms lifted on the desk in front
	Call_{LR}	One arm on the desk in front the other one at the ear
	Hunch	Sitting and leaning forward but tilting the head upwards
Stand	I	Standing upright with the arms hanging down
	T	Standing with both arms spread at a right angle
	X	Both arms spread at a right angle, both legs spread
	Call_{LR}	One arm hanging down the other one at the ear
	Walk_{LR}	One leg in front, arms in a natural swinging motion
	Fall_{LR}	Falling backwards, arms outstretched to regain balance

Table 8.1: All considered static postures with brief descriptions. Subscript LR mark that there are mirrored left and right versions of the corresponding postures, which results in a total of 14 postures.

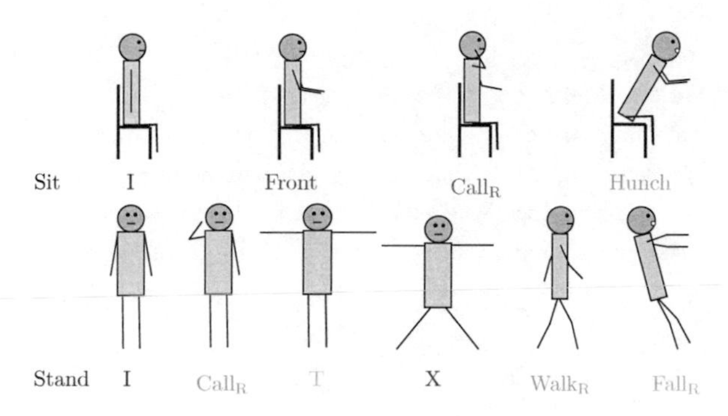

Figure 8.2: Illustration of the reference postures. For asymmetric postures only the right (R) version is shown.

are assumed to be unhealthy or dangerous, namely *Sit Hunch*, *Stand* $\textit{Fall}_{\text{L}}$ and *Stand* $\textit{Fall}_{\text{R}}$. Asymmetric postures with a left-right differentiation are denoted with indices *LR*, and both versions will be considered individually for the classification.

8.2 Noise Models

Next, we introduce different artificial noise models to efficiently study the sensitivity of the system and compare different system designs. To this end, the impedance vector $\mathbf{x}_p^0$ and the resulting feature vector $\mathbf{r}_p^0$ (cf. (7.3)) are subsequently impaired by various kinds of perturbations. These perturbations may result from the specific measurement device[1], model inaccuracies, or other practical uncertainties.

Additive White Gaussian Noise (AWGN) One type of perturbation that we consider in this posture recognition context is AWGN. It is not only the result of electron movement at the measurement device, but also a standard assumption given that many superimposing types of independent noise are present (cf. central limit theorem [139]). For the AWGN case, we model the noisy impedance vectors by

$$\mathbf{x}_p^{\mathrm{AWGN}} = \mathbf{x}_p^0 + \mathbf{w}^{\mathrm{AWGN}}, \quad \mathbf{w}^{\mathrm{AWGN}} \overset{\text{i.i.d.}}{\sim} \mathcal{CN}\left(\mathbf{0}_{N_\mathrm{R}\times 1}, (\sigma^{\mathrm{AWGN}})^2\,\mathbf{I}_{N_\mathrm{R}}\right), \tag{8.1}$$

with σ^{AWGN} denoting the noise standard deviation in ohms.

Lognormal Noise Another perturbation which we consider is lognormal noise. This type of noise is commonly used to model logarithmic quantities such as energies or concentrations. In a communications context, it is for example used to model the received power (in dB) over distance, which is referred to as the lognormal path loss model [140]. For our posture recognition analysis, it is considered to also cover multiplicative perturbations. For the impedance vector, we model it according to

$$\mathbf{x}_p^{\mathrm{logN}} = \mathrm{diag}\left(\mathbf{w}^{\mathrm{logN}}\right) \cdot \mathbf{x}_p^0, \quad \log(\mathbf{w}^{\mathrm{logN}}) \overset{\text{i.i.d.}}{\sim} \mathcal{N}\left(\mathbf{0}_{N_\mathrm{R}\times 1}, (\sigma^{\mathrm{logN}})^2\,\mathbf{I}_{N_\mathrm{R}}\right), \tag{8.2}$$

where σ^{logN} is the standard deviation of the observation's natural logarithm. This type of modeling introduces a correlation as it affects the real and imaginary parts of the impedance vector equally.

Inductive Perturbations Another typically considered source of perturbations in MI setups are the mutual inductances between all coil pairs. These may be affected by imperfect current densities on the wires and uncertainties of the coil geometry that cannot be calibrated for. Such changes may for example occur due to a stretching of the coils, which clearly affects their shape and thus also their characterization. Moreover, minor displacements of the coils (e.g. if they slip out of position) would also impact the mutual impedances in an irregular fashion. In order to roughly model the resulting impact of such errors on our impedance vector, we first consider the mutual inductance matrix $\mathbf{M}_p$ between all corresponding coils for a given posture p.

[1]Note that an experimental investigation of the observed measurement noise for the VNA used in Cpt. 9 can be found in Appendix C.

We multiply this matrix's individual off-diagonal elements $[\mathbf{M}_p]_{m,n}$ for $m \neq n$ by a multiplicative random variable according to

$$[\mathbf{M}_p^{\mathrm{indN}}]_{n,m} = [\mathbf{M}_p^{\mathrm{indN}}]_{m,n} = [\mathbf{M}_p]_{m,n} \cdot w^{\mathrm{indN}}, \quad w^{\mathrm{indN}} \overset{\mathrm{i.i.d.}}{\sim} \mathcal{TN}(1, (\sigma^{\mathrm{indN}})^2, [0,2]), \tag{8.3}$$

where we use $\mathcal{TN}(\cdot, \cdot, [a,b])$ for a truncated normal distribution in the interval $[a,b]$. Consequently, (8.3) leads to approximately, normally distributed but symmetric off-diagonal entries of the inductance matrix and σ^{indN} specifies the relative standard deviation with respect to each corresponding mean value. The perturbed mutual inductances are furthermore subjected to various subsequent mathematical operations, such as multiplications and inversions according to our MI model. We summarize all these operations by defining $\mathbf{x}_p^0 = \mathbf{g}(\mathbf{M}_p)$, where $\mathbf{g}(.)$ is the functional that combines the steps (2.14) and (7.1) to (7.3). Overall, we then obtain the perturbed impedance vector by applying

$$\mathbf{x}_p^{\mathrm{indN}} = \mathbf{g}(\mathbf{M}_p^{\mathrm{indN}}). \tag{8.4}$$

Joint Angle Perturbations Lastly, we also want to consider AWGN on all the joint angles which connect the torso and limbs of the human body models from Sec. 8.1. This consideration is important as slight variations of the original true posture will inevitably be present in a real scenario. By explicitly modeling this type of noise, we want to understand how severely minor posture variations translate into a loss of the classification accuracy. Defining $\mathbf{\Phi}_p$ as a vector containing all the true rotatable joint angles (cf. Sec. 8.1) for a given posture, each element $[\mathbf{\Phi}_p]_k$ with $k = 1, \ldots, 22$ is affected as

$$[\mathbf{\Phi}_p^{\mathrm{Joints}}]_k = [\mathbf{\Phi}_p]_k + w^{\mathrm{Joints}}, \quad w^{\mathrm{Joints}} \overset{\mathrm{i.i.d.}}{\sim} \mathcal{TN}(0, (\sigma^{\mathrm{Joints}})^2, [-180, 180]). \tag{8.5}$$

This leads to w^{Joints} being approximately normally distributed with zero-mean in the truncated interval $[-180, 180]$. The standard deviation σ^{Joints} is always stated in degree. Since the mutual inductance matrix varies for different realizations of the joint angles (and hence postures), the overall noisy impedance impedance vector follows as

$$\mathbf{x}_p^{\mathrm{Joints}} = \mathbf{g}\left(\mathbf{M}(\mathbf{\Phi}_p^{\mathrm{Joints}})\right). \tag{8.6}$$

By modeling the perturbations in the above manner, any joint on the limbs (e.g. the elbow joint) is subject to cumulative perturbations as the individual errors on all connecting joints lead to a compound effect. The limb joints may thus end up with a higher overall variation. Furthermore, we constrain all perturbed postures such that an intersection of coils becomes impossible. Note that this modeling does not respect anatomical constraints of the joints.

Example of Perturbed Impedance Measurements An illustrative comparison of the different impacts on the real and imaginary parts of the impedance vector is shown in Fig. 8.3. This result is based on a specific placement of four anchors and nine passive tags (cf. Sec. 8.3) on the human model Martin with the fixed Stand $\mathrm{Call}_{\mathrm{R}}$ posture. The yellow marker represents the noiseless or unperturbed case, which shows how drastically the input impedances of all anchors are detuned from the original $R^{\mathrm{ref}} = 50\,\Omega$ due to the combined impact of passive tags. This significant detuning of the input impedances is important, since a posture classification would be impossible if these changes were imperceptible. The red and blue markers show how the different types of perturbations affect the corresponding impedance vectors for different standard deviations (or noise levels) of the corresponding perturbation. As is apparent from (8.1), we observe typical Gaussian point clouds for the AWGN case, whereas the Lognormal case by construction leads to straight lines which reflect the multiplicative attenuation or amplification of the impedances. For the perturbations of the mutual inductances and joints, we see a more intricate behavior for each individual anchor and a strong correlation of real and imaginary parts. Ultimately, the individual anchors are thus not only differently affected by the passive tags, but also by the noise realizations. This means that each anchor may provide different posture information and noise characteristics, so using multiple anchors may be beneficial for the classification.

Fig. 8.4 shows a similar scenario, in which the input impedances of one single anchor are given for all 14 different postures. The yellow markers again represent the noiseless input impedances for each different posture, whereas the red and blue markers show noise realizations with different standard deviations. In contrast to the previous plot, this illustration already gives some intuition on the classification capabilities of our system. That is, we see that the postures can easily be separated into two distinct clusters (upper right and lower left), which indeed corresponds to all standing and all sitting postures, respectively. However, we also see that within these clusters, some postures overlap even for small noise realizations. This means that if we were to only rely on this single anchor, then a classification of similar postures fails quickly for higher noise levels.

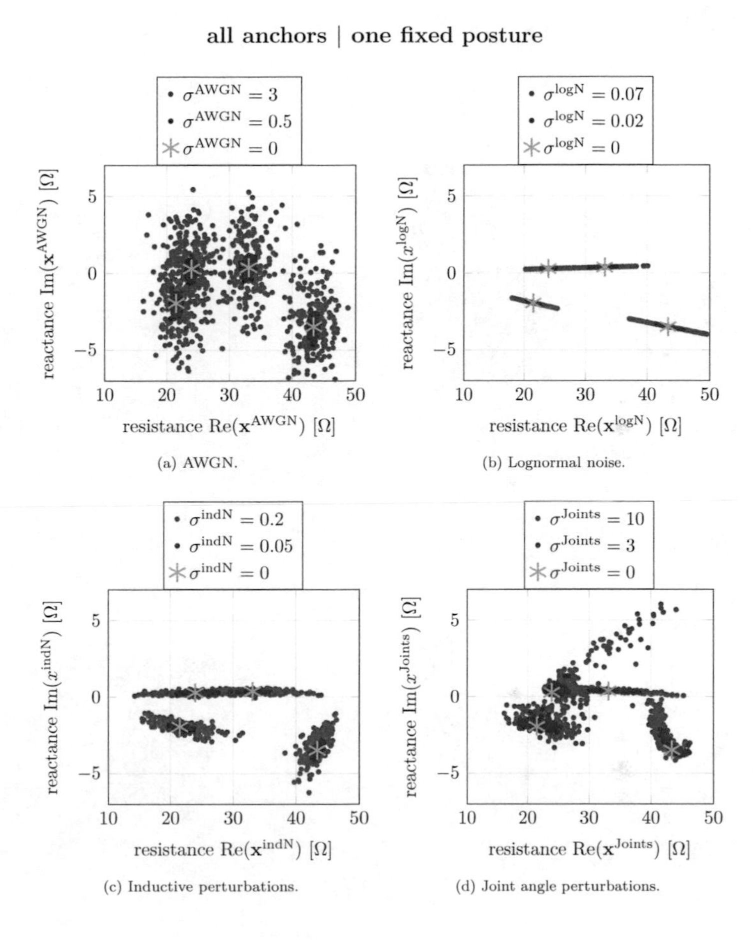

(a) AWGN.

(b) Lognormal noise.

(c) Inductive perturbations.

(d) Joint angle perturbations.

Figure 8.3: Exemplary perturbed input impedances of four anchors for a single fixed posture. The input impedances are displayed in the complex plane for AWGN (a), lognormal noise (b), inductive perturbations (c) and joint angle perturbations (d).

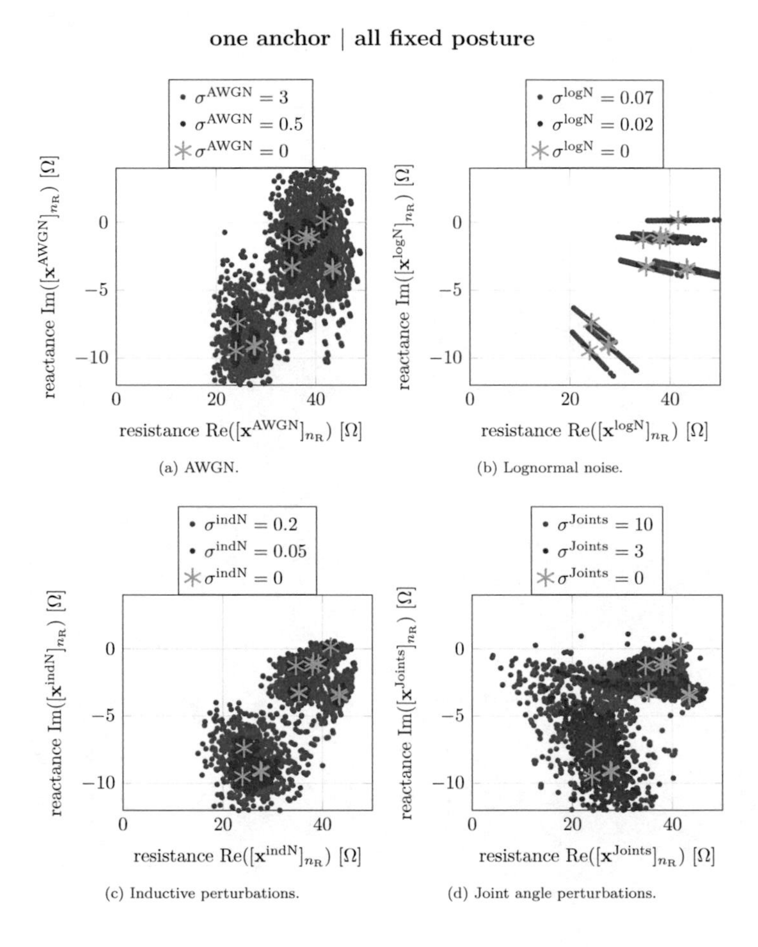

Figure 8.4: Exemplary perturbed input impedances of a single anchor for 14 different posture. The input impedances are displayed in the complex plane for AWGN (a), lognormal noise (b), inductive perturbation (c) and joint angle perturbation (d).

8.3 Coil Types and Placement

In this section, we introduce two different coil designs which we place on the body models to compare their suitability for posture recognition. The coil designs and the coil placement should generally comply with recommended design considerations of Sec. 7.2 in order for the system to work well. However, in a first instance they also have to satisfy the following practical criteria which may object some of the earlier considerations.

1. The coils have to fit comfortably on the human body or they have to be implementable into the clothing. This requirement restricts the coil geometry and e.g. complicates the use of three-axis coils.

2. The anchor placement should be centralized and restricted to one piece of clothing, such that a single integrated measurement device suffices to obtain all impedance measurements. Further, they should be placed such that the distance to the passive tags is not unnecessarily large.

3. The coils need to couple sufficiently over the entire range of the human body. Yet, the coils and their additional circuitry also need to be inexpensive, which complicates the design of high-quality coils. It may hence be beneficial to compensate cheaper materials by using larger coils, as long as they can be still be worn comfortably.

4. The anchor coils have to be designed in such a way, that their self-resonance frequencies are significantly higher than the operating frequency. If this criterion is not met, open-circuited (deactivated) anchors would cause relevant interference with the impedance measurement of each actively measuring anchor. This disturbance might overshadow the detuning caused by the passive tags.

With these criteria in mind, we decided to focus on two different coil designs: (i) simple solenoid coils that surround a body part and (ii) flat spider web coils, which are placed evenly on a body part's surface. The coil designs and their placements are shown in Fig. 8.5a and Fig. 8.5b, where they are mounted on the Martin model. The general coil and simulation parameters are given in Tab. 8.2. Other coil parameters that vary for different coils can be found in Tab. 8.3 and Tab. 8.4 for the Martin model. The quality factors are stated for $f^{\mathrm{res}} = f^{\mathrm{des}}$.

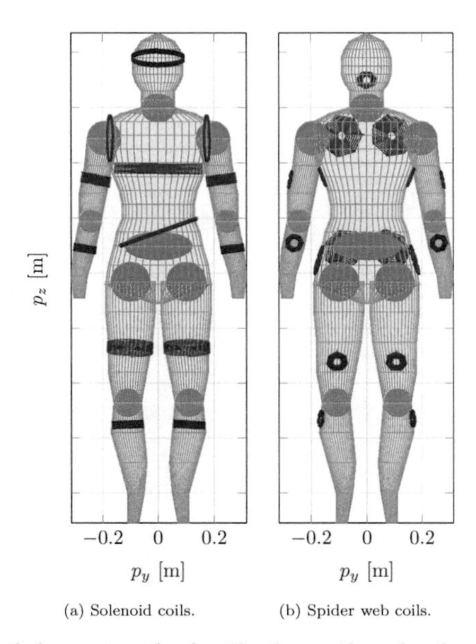

(a) Solenoid coils.

(b) Spider web coils.

Figure 8.5: Suggested coil placements with solenoid coils or spider web coils on the male human body model Martin. The anchors (blue) are placed on the torso and passive tags (red) on the limbs. The spider web configuration has twice the number of anchor coils.

wire diameter	D^{wire}	254.6 µm
design frequency	f^{des}	500 kHz
op. frequency	f	500 kHz
conductivity	σ^{wire}	59.6MS/m
rel. permittivity	ϵ_{r}	1
rel. permeability	μ_{r}	1

(a) Specified parameters.

wavelength	λ	600 m

(b) Resulting parameters.

Table 8.2: Simulation parameters and resulting quantities.

Parameter → Coil ↓		D^{coil} [cm]	N^{coil} [1]	H^{coil} [cm]	R [Ω]	L [µH]	C [pF]	f^{self} [MHz]	Q [1]
Receiving anchors	n_{R}								
$\text{Shoulder}_{\text{LR}}$	1,2	17	5	1	1	9.7	21.8	11	30.5
Hip	3	29.4	10	1.5	3.4	68.7	41	3	63.5
Torso	4	31.2	10	4	3.5	56.8	26.4	4.1	51
Tags	n_{T}								
Calf_{LR}	1,2	11.2	10	3	1.9	35.1	6.9	10.2	58
Thigh_{LR}	3,4	16.6	10	4	1.9	24.5	10.6	9.9	40.5
$\text{Forearm}_{\text{LR}}$	5,6	8	15	3	1.4	21.8	4.4	16.3	48.9
$\text{Upper arm}_{\text{LR}}$	7,8	10.8	10	4	1.2	13.5	5.9	17.8	35.3
Head	9	19	10	2	2.2	36.7	17.8	6.2	52.4

Table 8.3: All solenoid coils and their parameters for the Martin model as illustrated in Fig. 8.5. All values are rounded to single decimals.

Parameter → Coil ↓		D^{coil} [cm]	N^{coil} [1]	H^{coil} [cm]	R [Ω]	L [µH]	C [pF]	f^{self} [MHz]	Q [1]
Receiving anchors	n_{R}								
Lat_{LR}	1,2	5.7	15	2.5	1	11.2	3	27.5	35.2
Torso_{LR}	3,4	10.3	15	6.5	1.6	16.8	5	17.4	33
Hip_{LR}	5,6	8	15	4.5	1.3	13.8	4	21.5	33.3
$\text{Bottom}_{\text{LR}}$	7,8	10.3	15	6.5	1.6	16.8	5	17.4	33
Passive tags	n_{T}								
Calf_{LR}	1,2	5.2	10	2	0.6	5	2.8	42.7	26.2
Thigh_{LR}	3,4	5.7	15	2.5	1	11.2	3	27.5	35.2
$\text{Forearm}_{\text{LR}}$	5,6	5.2	10	2	0.6	5	2.8	42.7	26.2
$\text{Upper arm}_{\text{LR}}$	7,8	5.2	10	2	0.6	5	2.8	42.7	26.2
Head	9	5.2	10	2	0.6	5	2.8	42.7	26.2

Table 8.4: All spider web coils and their parameters for the Martin model as illustrated in Fig. 8.5. All values are rounded to single decimals. Note that the effective overall diameter of a spider web coil is given by the root mean square of the diameters of each coil turn.

The different impacts of the two coil designs, are analyzed for the Martin model in Fig. 8.6. That is, for each posture p we have one unperturbed feature vector $\mathbf{r}_p^0$ of dimension $2N_\mathrm{R} \times 1$ (cf. (7.3)). Both Fig. 8.6a (solenoid) and Fig. 8.6b (spider web) show the first and second components of the corresponding **P**rincipal **C**omponent **A**nalysis (PCA) [141] over all $N_p = 14$ postures. These primary principal components are the two orthogonal linear combinations of the anchor observations, which explain most of the overall variance that results from taking on different postures. Each plotted point corresponds to the feature vector $\mathbf{r}_p^0$ of a given posture that is projected from the full feature space onto these two principal dimensions. The explained variance of the principal components represents the main information available to separate the different postures. It can be seen from Fig. 8.6a that even with only two components, a separation of most postures is easily possible with the solenoid setup. The differentiation between the asymmetric left and right postures on the other hand shows to be problematic. This is partially caused by the fact that both principle components dependent mainly on the hip and torso anchor coils. Yet, for these two coils the left and right postures lead to almost identical impedance observations. Looking at the analogous PCA of the spider web setup in Fig. 8.6b, it is clear that the separation of the different postures is more complicated as some postures cluster more severely, namely Sit I and Stand I, as well as Sit Call$_\mathrm{L}$, Sit Call$_\mathrm{R}$, Sit Hunch, Sit Front, Stand X, Stand T, Stand Fall$_\mathrm{L}$ and Stand Fall$_\mathrm{R}$. We presume that this clustering is the result of (i) a weaker coupling between the spider web coils due to the smaller coil sizes that are required for a flat on-body placement and (ii) having overall lower quality factors. In Fig. 8.6c and Fig. 8.6d we show the corresponding Scree plots [142] of both PCAs. For the solenoid setup, it can be seen that approximately five principal components are necessary to explain the variance. In contrast, the observation variance for the spider web setup can be approximately explained only based on two different principal components, so the observations are more linearly dependent (i.e. there is more redundancy in our impedance observations).

However, it is not only important how many independent observations are required to explain the posture variance, but also how well this variance separates all individual postures. To this end, the Fig. 8.6e shows the numerical values of the minimum and median inter-posture distances in the full feature space for both setups. For a posture p, the minimum inter-posture distance is determined as $d_p^\mathrm{min} = \min_{q,q\neq p} ||\mathbf{r}_p^0 - \mathbf{r}_q^0||$ and the median inter-posture distance follows as $d_p^\mathrm{med} = \mathrm{median}_{q,q\neq p} ||\mathbf{r}_p^0 - \mathbf{r}_q^0||$, where $q \neq p$ denotes all postures other than p. The minimum distances are an important measure of the noise sensitivity for each coil design. For the Gaussian case and individual

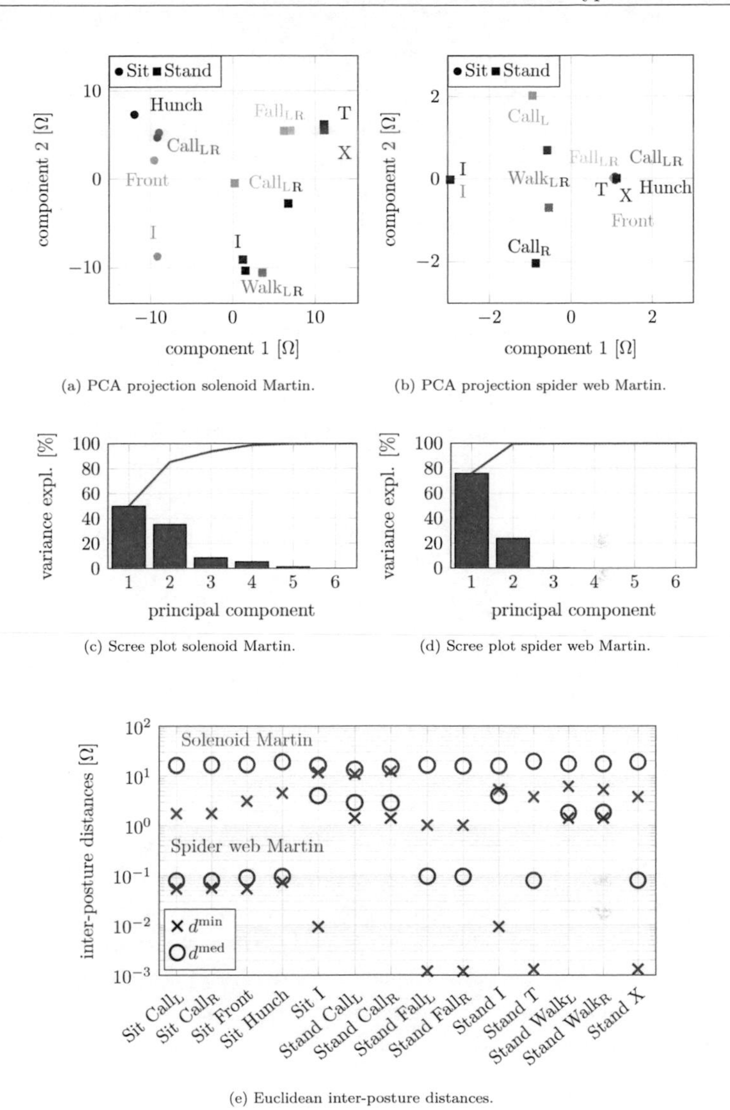

(a) PCA projection solenoid Martin.

(b) PCA projection spider web Martin.

(c) Scree plot solenoid Martin.

(d) Scree plot spider web Martin.

(e) Euclidean inter-posture distances.

Figure 8.6: The first row shows the two primary principal components for the solenoid (left) and the spider web coil (right) setup. The second row shows the Scree plots for the first six principal components of both configurations. The last row shows the minimum and median Euclidean distances of the noiseless feature vector of a given posture with respect to all other postures.

pairwise decisions, these minimum distance can further be translated into pairwise error probabilities (i.e. the probability of pairwise misclassification) via Q-function. For the other perturbations this relationship is not as straightforward. Nevertheless, higher minimum and median distances between postures are generally associated with lower error probabilities. From Fig. 8.6e it is clear that both the minimum and the median distances of the solenoid setup are significantly larger than those of the spider web setup. That is, for each posture p the minimum distance of the solenoid setup is larger than the corresponding median distance of the spider web setup. In a direct comparison, we thus expect the solenoid setup to offer a superior separability and fewer misclassifications. We further deduce that the chosen spider web coil design and placement may be unsuited for the classification of our considered postures, since most of the minimum distances are in the mΩ regime and will easily be bridged by noise or model uncertainties.

Interference from Neighboring Systems With regard to massive deployment, we also evaluated the impact of multiple systems being operated in close proximity. To this end, we looked at a scenario in which two full body Martin models with the solenoid coil setup are separated by a distance d^{distort} in p_x direction (cf. Fig. 8.5). One of the models remains in the *Stand I* posture while all of its coils, even the anchor coils, are resonantly loaded. The other model operates as intended and measures its own and now distorted anchor impedances $Z^{\mathrm{in,distort}}_{\tilde{\mathrm{R}},n_{\mathrm{R}}}$ sequentially for all different postures. We next look at the relative error $\eta^{\mathrm{distort}}_{n_{\mathrm{R}}} = \left|Z^{\mathrm{in}}_{\tilde{\mathrm{R}},n_{\mathrm{R}}} - Z^{\mathrm{in,distort}}_{\tilde{\mathrm{R}},n_{\mathrm{R}}}\right| \cdot \left|Z^{\mathrm{in}}_{\tilde{\mathrm{R}},n_{\mathrm{R}}}\right|^{-1}$ caused by this neighboring system if other perturbations are fully omitted. In Fig. 8.7, the maximally observed relative error (over all p) is visualized for different distances d^{distort}. We see that the influence of the second model decreases quickly, as expected. At a distance of $d^{\mathrm{distort}} = 1.2\,\mathrm{m}$, the worst overall impact only corresponds to a distortion of the true impedance magnitude by roughly 0.1%. This quick decay of the interference is a direct result of the strong path loss of the magnetic near field and is sometimes also referred to as magnetic bubble [29]. As the resulting impact is significantly smaller than the perturbations considered in Sec. 8.5 , we omit its impact altogether and assume a sufficient distance to possibly distorting materials.

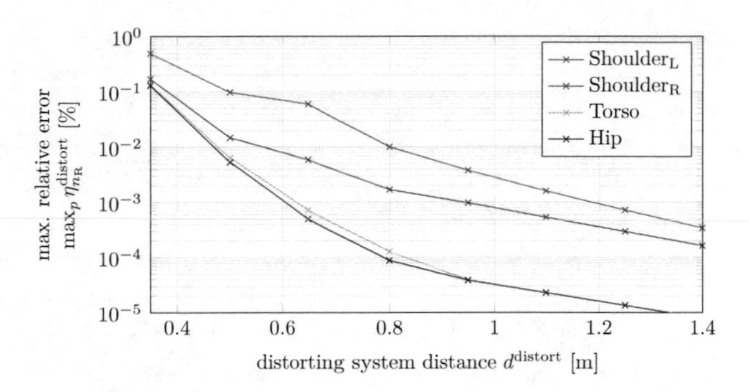

Figure 8.7: Maximum (over all p) relative error at $f = 500\,\mathrm{kHz}$ as a result of a neighboring MI posture recognition system with resonantly loaded anchors at distance of d^{distort}.

8.4 Data Set Generation and Classifiers

The previous models are used to generate a dataset that can be used for training and testing purposes of various classifiers. More precisely, for each full parametrization that specifies body model, coil design, posture, noise model and noise level, a perturbed impedance vector $\mathbf{x}_p$ is generated according to (8.1) to (8.6). This complex-valued impedance vector is further partitioned into the corresponding real-valued realization of the perturbed feature vector $\mathbf{r}_p$. We generate 1000 of these $2N_{\mathrm{R}} \times 1$-dimensional feature vectors per parametrization. For each parametrization, the total dataset is further split into a training part (70 %, 700 feature vectors) and a testing part (30 %, 300 feature vectors). Note that the postures in either part are still equiprobable. This dataset separation is furthermore fixed, meaning that no feature vectors of the test part are ever considered by a training process.

Using the training part, we train any type of classifier via five-fold cross validation to mitigate overfitting. That is, we use random 70 percent of the training part for the learning process of a single classifier and the remaining 30 percent for the corresponding validation of the same classifier. If we evaluate the performance of the training part, we solely use the average misclassification rate as performance metric. If we instead want to examine the actual performance, we evaluate the classifier's accuracy (the total ratio of correct classifications, i.e. accuracy = 1 − misclass. rate) as performance metric on the testing part. Note that if not stated otherwise, the classifiers are always trained with the same body model, coil design, noise model and noise level that they are tested on.

8.4.1 Classifier Comparison During Validation

In the following subsection, we investigate the performance of three different types of classifiers, namely the **k**-**N**earest **N**eighbors (kNN) algorithm, **S**upport **V**ector **M**achines (SVMs) and **M**ulti**L**ayer **P**erceptrons (MLPs). For each of these classifiers, we compare the impact of different key hyperparameters on the performance during validation. The other hyperparameters are determined via random search. In this preliminary study we only use the Martin model, solenoid coils and AWGN.

kNN As one of the simpler and more established approaches, we employ a kNN classifier [141]. It uses the Euclidean distance between neighbors, weights all distances equally, searches for the neighbors via Kd-tree and breaks ties randomly. In order to decide on the number of neighbors that should be considered, Fig. 8.8a shows a colorplot that highlights the impact of an increasing k on the number of misclassifications during the cross-validation step. We see that for all considered noise levels, the improvements diminish drastically for an increasing number of neighbors k. Even for the highest noise level, there is no relevant improvement after $k = 15$ and we thus fix this choice as our operating point. This result was also validated to work well for the spider web setup and the Fiona model (not shown).

SVM We also consider a SVM that performs multi-class decisions with the One-vs-One approach, i.e. by applying a series of binary decisions [141]. The final multi-class decision is made based on the class with the most wins after performing all possible binary decisions. The SVM also uses the kernel trick in order to allow for quick non-linear decisions in the feature space [141]. In Fig. 8.8b we compare the performance of SVMs with different kernels during the cross-validation. More precisely, we analyze the performance for a linear SVM, a polynomial kernel of order 3 and a **R**adial **B**asis **F**unction (RBF) kernel.

MLP Lastly, we also analyze the performance of feed forward MLPs for the classification [141]. For all our MLPs, we use a hyperbolic tangent sigmoid function at each neuron, and decide for the posture which corresponds to the maximum output of the final softmax layer. In Fig. 8.8c, we show the misclassification rate during validation for multiple noise levels. The compared MLPs only differ with respect to the number of hidden layers used and neurons per layer, which are indicated by the legend. The blue line for example corresponds to a MLP with only two hidden layers, which each comprise 28 neurons. Overall, Fig. 8.8c shows no performance difference as a result of the network size. To mitigate possible overfitting, we hence continue with the simplest structure MLP structure, namely the one with only two hidden layers that each contain 28 neurons.

Comparing all three types of classifiers, we find a similar performance behavior. That is, all classifiers start with a perfect misclassification rate of approximately 0 for low standard deviations σ^{AWGN}, which increases steadily to roughly 60 % at $\sigma^{\mathrm{AWGN}} = 7\,\Omega$. Ultimately, we assume that at least with AWGN and our the low-dimensional feature space, no intricate decision boundaries are required. As a result, even simple classification algorithms with few parameters suffice. This behavior may however differ for other types of perturbations.

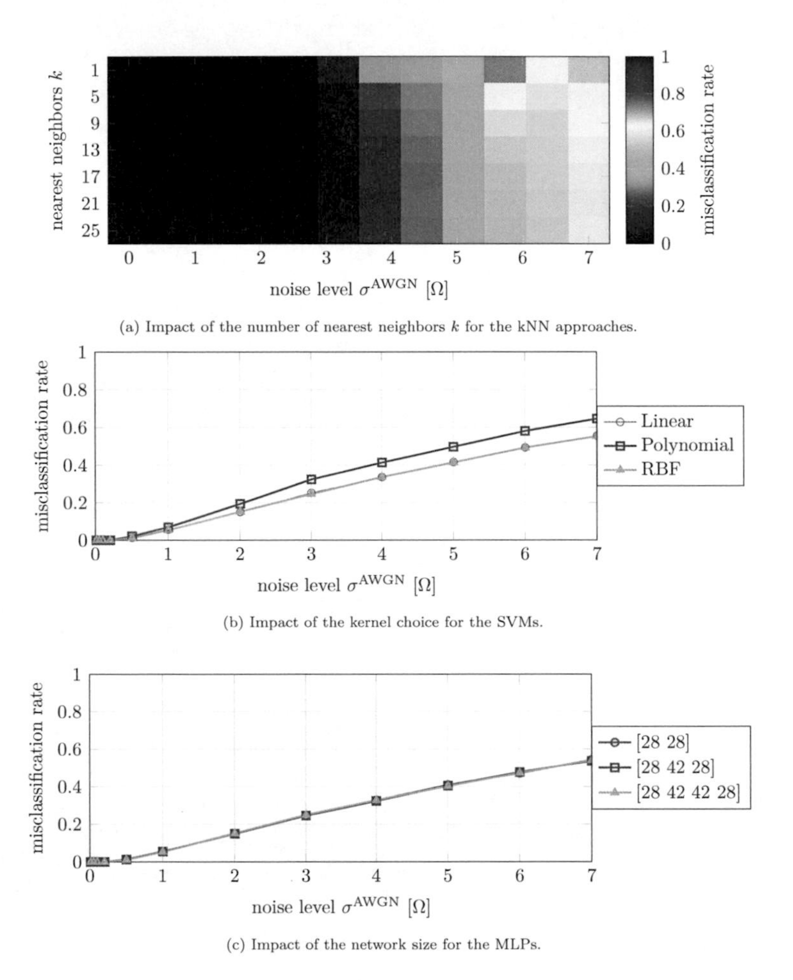

(a) Impact of the number of nearest neighbors k for the kNN approaches.

(b) Impact of the kernel choice for the SVMs.

(c) Impact of the network size for the MLPs.

Figure 8.8: Impact comparison of typical hyperparameters on the misclassification rate during the validation process for three different classifiers using the Martin model with solenoid coils and AWGN.

8.5 Performance Evaluation

One remaining issue is the fundamental difference between the four types of perturbations, which complicates a direct comparison. It is thus helpful to establish a common basis when comparing their impact. For our analyses, we hence choose their standard deviations such that they cause the same resulting expected deviations on the impedances vectors, i.e. we require

$$\begin{aligned}(\sigma^{\mathrm{AWGN}})^2 = \mathrm{E}\left[\|\mathbf{x}_p^{\mathrm{AWGN}}(\sigma^{\mathrm{AWGN}}) - \mathbf{x}_p^0\|^2\right] &\overset{!}{\approx} \mathrm{E}\left[\|\mathbf{x}_p^{\mathrm{logN}}(\sigma^{\mathrm{logN}}) - \mathbf{x}_p^0\|^2\right] \\ \overset{!}{\approx} \mathrm{E}\left[\|\mathbf{x}_p^{\mathrm{indN}}(\sigma^{\mathrm{indN}}) - \mathbf{x}_p^0\|^2\right] &\overset{!}{\approx} \mathrm{E}\left[\|\mathbf{x}_p^{\mathrm{Joints}}(\sigma^{\mathrm{Joints}}) - \mathbf{x}_p^0\|^2\right],\end{aligned}$$

with the expectation being taken over all postures and noisy realizations for an otherwise specified parametrization. We performed this standardization for the maximum values of the corresponding standard deviations and found that $\sigma^{\mathrm{AWGN}} = 7\,\Omega$, $\sigma^{\mathrm{logN}} = 0.4$, $\sigma^{\mathrm{indN}} = 0.4$, and $\sigma^{\mathrm{Joints}} = 20°$ all lead to the roughly same impedance deviations. This clearly also holds for $\sigma^{\mathrm{AWGN}} = 0\,\Omega$, $\sigma^{\mathrm{logN}} = 0$, $\sigma^{\mathrm{indN}} = 0$, and $\sigma^{\mathrm{Joints}} = 0°$. All other considered standard deviations in-between are not necessarily of equivalent impact.

8.5.1 Comparison of Classifiers

In Fig. 8.9 the classifiers accuracy is compared on the testing data for these fixed noise ranges. For AWGN, all classifiers yield a similar accuracy that agrees with the previous results of Fig. 8.8. Moreover, the performance decline is already visible with low noise contributions, which means that some decision boundaries lie close to the true values. For the lognormal and inductive perturbations, it is clear that the SVMs and MLPs drastically outperform the kNN approach. Overall, we see that at least for the solenoid Martin case, the classification with the SVM is resistant against each noise type and maintains a 90 percent accuracy if the standard deviations are lower than $\sigma^{\mathrm{AWGN}} = 1.5\,\Omega$, $\sigma^{\mathrm{logN}} = 0.35$, $\sigma^{\mathrm{indN}} = 0.21$, and $\sigma^{\mathrm{Joints}} = 8°$, respectively. In the following, we will use the SVM as classifier of choice due to its high accuracy and low computational complexity. Further, while these preliminary results seem promising, it is unlikely that only one type of perturbation is present in reality. Minor variations of the users posture may for example occur independently of additional thermal noise. In Sec. 8.5.2 we thus examine how a combination of multiple noise types affects the classification.

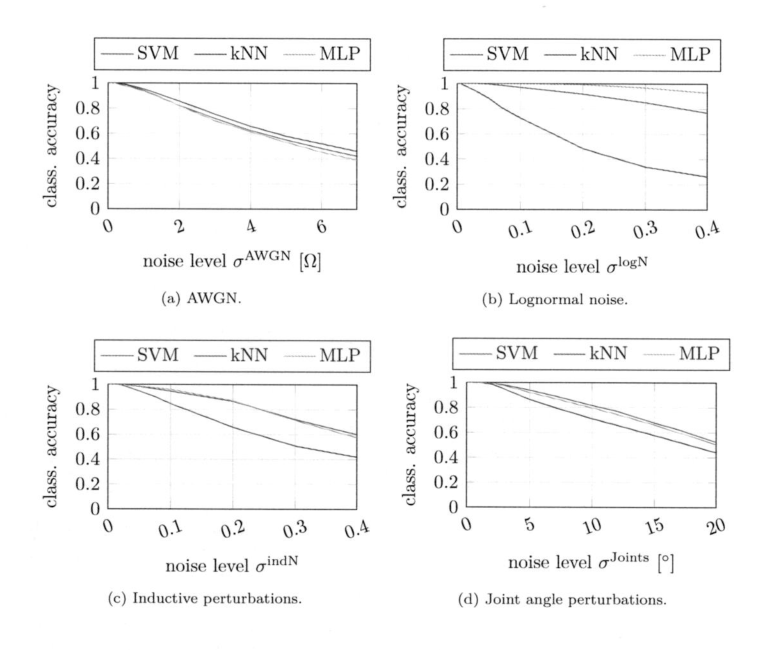

(a) AWGN.

(b) Lognormal noise.

(c) Inductive perturbations.

(d) Joint angle perturbations.

Figure 8.9: Classification accuracy of the kNN, SVM, and MLP classifiers for different noise models and noise levels. The results are obtained using solenoid coils on the Martin model.

8.5.2 Mixed Perturbation

Next, we look at a mixed perturbation which combines AWGN and joint angle perturbations. That is, the corresponding complex impedance vectors of this mixed perturbation have the form

$$\mathbf{x}_p^{\text{mixed}} = \mathbf{x}_p^{\text{Joints}} + \mathbf{w}^{\text{AWGN}} . \tag{8.7}$$

For this mixed perturbation, Fig. 8.10 shows the classification accuracy that is obtained on the corresponding testing data when using the solenoid Martin case in combination with a SVM. The two axes show the increasing noise standard deviations, whereas the color represents the classification accuracy. The dashed white lines mark the equipotential lines. These lines highlight that for a given σ^{AWGN} and a small $\sigma^{\text{Joints}} < 2°$, the classification performance remains approximately constant, so minor posture changes are not detrimental to the system performance. For higher standard deviations σ^{Joints}, the equipotential lines exhibit a linear behavior, i.e. they correspond to points $(\sigma^{\text{Joints}}, \sigma^{\text{AWGN}})$ that approximately satisfy $a\,\sigma^{\text{Joints}} + b\,\sigma^{\text{AWGN}} = c$ with $a, b, c \in \mathbb{R}$. The plot further indicates how stronger standard deviations of a single noise type may be compensated for by reducing the standard deviation of the other noise type. However, this compensation potential is limited by practical constraints, e.g. the cost of the measurement device or the available measurement duration. In order to decide on the overall feasibility, it is thus necessary to first identify realistic operating points.

For the AWGN, we did so by evaluating different suitable measurement methods and by comparing officially stated tolerances for impedance meters and for low-cost measurement boards that operate in a comparable regime [137,138]. Overall, we hence expect a measurement tolerance which leads to 99% of errors being within $\pm 1\,\%$ of the true impedance value. Yet, such percental errors cannot be implemented when using the generic AWGN noise model with identical standard deviations on all impedance measurements. We alternatively use a pessimistic workaround and simply choose the standard deviation corresponding to the highest impedance value observed for all anchors and postures (cf. Fig. 8.3 and Fig. 8.4), which is roughly $50\,\Omega$. Consequently, for 99% of errors to be in the range of $[-0.5\,\Omega, 0.5\,\Omega]$, the standard deviation under a Gaussian assumption follows via quantile function [143] as $\sigma^{\text{AWGN}} = \frac{0.5\,\Omega}{(\sqrt{2}\text{erfc}^{-1}(0.01)}} \approx 0.2\,\Omega$. However, since we choose a pessimistic approach, lower values of σ^{AWGN} may be acceptable as well. To further assess reasonable deviations for the joint angle perturbations,

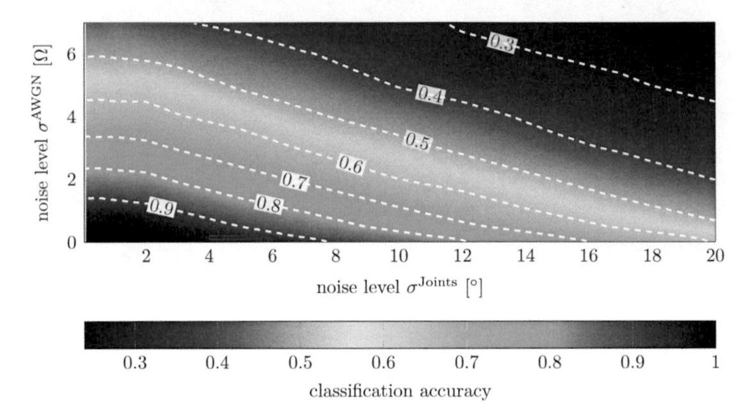

Figure 8.10: Colorplot of the obtained classification accuracy via SVM for the solenoid Martin model when being subjected to both AWGN and joint angle perturbations. The dashed white lines represent the equipotential lines of the accuracy. The red rectangle highlights reasonable operating points.

we instructed three test subjects to re-enact all postures (as would be done during a calibration) and measured the variations of their joints compared to the corresponding true joint angles. This led to deviations that translated to $\sigma^{\text{Joints}} \in [4°, 6°]$ on average. These practical considerations led us to believe that reasonable operating points lie within the highlighted red rectangle.

For the subsequent analysis, we will focus on a single operating point with $\sigma^{\text{AWGN}} = 0.2\,\Omega$ and $\sigma^{\text{Joints}} = 5°$. The confusion charts of all body models and coil designs for this operating point are displayed in Fig. 8.11. For both body models wearing the solenoid coils, we see that the main classification errors are the result of left-right confusions for the *Stand Fall* and *Sit Call* postures. We assume that these errors are so frequent because left-right postures lead to the same detuning effect for the important large torso coil and to very similar detuning effects for the coil on the hip. The decisions thus have to rely on the observations of the shoulder coils, which are generally less resistant to noise because they observe smaller impedance variations. Unfortunately, these shoulder coils are strongly misaligned to the passive tags on the arms for the fall-related postures, which explains the even worse discriminability. Nevertheless, if these postures are of particular importance this issue can be addressed by an improved anchor coil placement, e.g. by moving the upper end of the shoulder coils closer to the neck and letting them cross the chest (cf. Fig. 8.5a). Moreover, when comparing the unhealthy postures (*Sit Hunch* and both *Stand Fall* postures) to the healthy ones, we

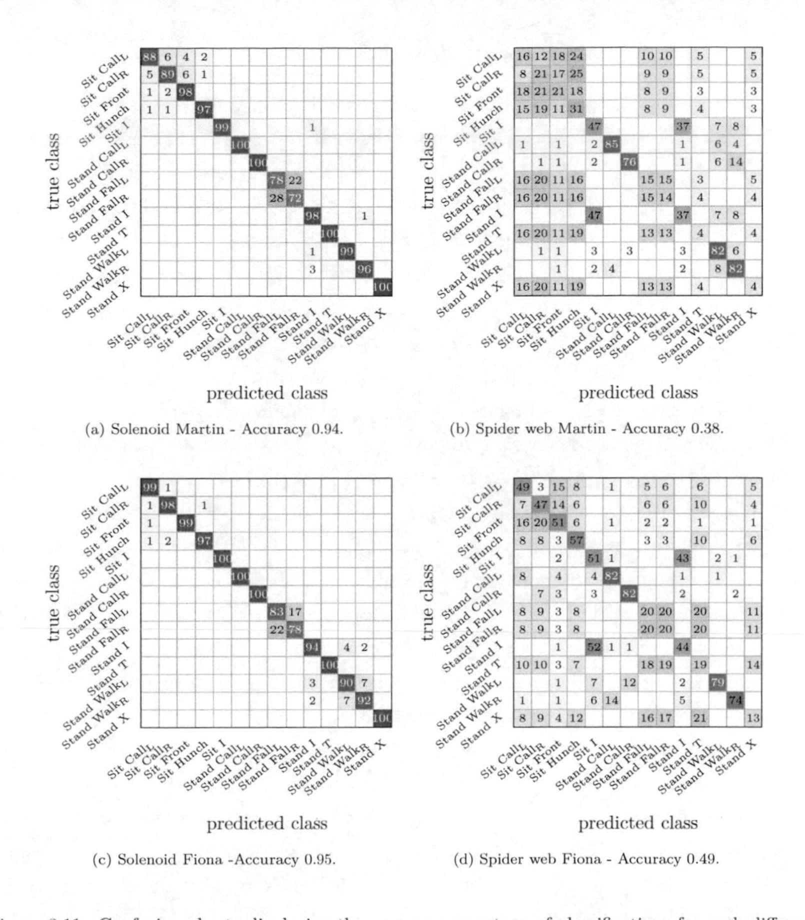

(a) Solenoid Martin - Accuracy 0.94.

(b) Spider web Martin - Accuracy 0.38.

(c) Solenoid Fiona -Accuracy 0.95.

(d) Spider web Fiona - Accuracy 0.49.

Figure 8.11: Confusion charts displaying the average percentage of classifications for each different posture based on evaluations on the unknown testing data. All entries are rounded to integers and the sum of each row and column might hence be unequal to 100.

see that false classifications between these two superclasses only occur in about 0.9 % of all cases. With the spider web coils a significantly degraded accuracy of less than 50 % is observed for either body model. This was already expected when we compared the two coil designs in Fig. 8.6 and is a result of the weak MI links between the spider web coils. The confusion charts thus further emphasizes how important large coils or sufficiently high quality factors are for posture recognition with passive MI tags.

8.5.3 Mismatch of Training and Test Data

In a last step, we also want to examine the generalizability of the training data for the two different body models with solenoid coils at the previously chosen operating point ($\sigma^{\text{AWGN}} = 0.2\,\Omega$, $\sigma^{\text{Joints}} = 5°$). That is, we look at the classification accuracy that is obtained if the SVMs trained on one body model are evaluated on testing data corresponding to the other body model. In Fig. 8.12 the corresponding confusion charts for this mismatched case with out-of-domain data are displayed. Here, Fig. 8.12a shows the case where the classifiers are trained using the solenoid Fiona data but are tested on the solenoid Martin data. Fig. 8.12b represents the opposite scenario. We clearly see that most classifications result in the same few classes with the overall accuracy being comparable to that of a random guess, namely $\frac{1}{14}$. As almost all decisions are affected, we deduct that this phenomenon occurs even in case of insignificant noise realizations. We hence infer that there is a severe shift of all clusters of feature vectors between both body models. More precisely, there is a severe domain shift (also distributional shift), so a relevant mismatch between the distributions of the training and testing data [144]. Without any domain adaption techniques, the learned decision boundaries are thus not suited to deal with the different input distributions and a generalization is not possible. In contrast, we can also train the classifiers using a mixed training set which comprises the training data of both body models. These more general classifiers are then evaluated on the separate testing data of both body models in Fig. 8.12c and Fig. 8.12d, respectively. The results are on par with our previous findings from Fig. 8.11, yielding an accuracy of roughly 93 % for either body model. This numerical evaluation indicates that even when looking at both body models simultaneously, there are no drastically overlapping posture clusters in the feature space. The bad results from the mismatched cases were hence indeed the result of a domain shift and the associated insufficient training data.

8.6 Conclusions

We investigated the usability of the proposed topology classification system for posture recognition. We introduced two different human body models and two different coil designs, including flat spider web coils and large solenoid coils. We found that only the large solenoid coils, which each fully surround a body part, were capable of causing a relevant impedance detuning for either body model. Additionally, we compared different supervised classifiers, namely kNN, SVMs, and MLPs, which all exhibited

incorrect training data (mismatch)

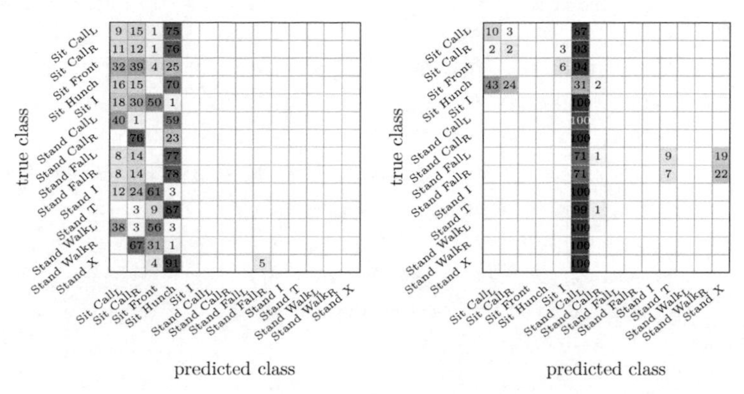

(a) Trained Fiona | Tested Martin: Accuracy 0.067. (b) Trained Martin | Tested Fiona: Accuracy 0.08.

mixed training data

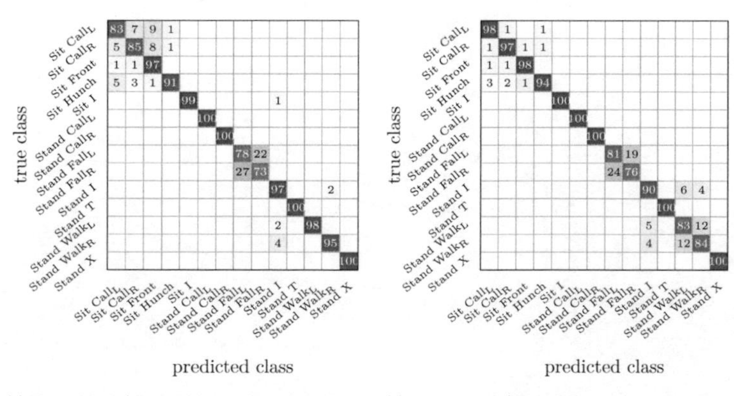

(c) Trained both | Tested Martin: Accuracy 0.926. (d) Trained both | Tested Fiona: Accuracy 0.93.

Figure 8.12: Confusion charts displaying the average percentage of classifications for each different posture based on different combinations of training and testing data. All entries are rounded to integers and the sum of each row and column might hence be unequal to 100.

a similar classification accuracy. This led us to believe that the classification within the eight-dimensional feature space does not require highly complex classifiers. Simple classifiers with few parameters that are less likely to overfit are hence preferable. We further characterized the robustness of the system against four different types of perturbations and identified reasonable operating points for a mixture of the different perturbations. For these reasonable operating points, the SVM obtained a classification accuracy of more than 90 % for both body models. This high accuracy is adequate for the applications motivated in Sec. 1.3. Nevertheless, this accuracy can only be obtained if the training and testing data come from the same body model. If there is a mismatch, i.e. if the classifiers are trained exclusively on measurements from the female body model and tested on measurements from the male body model (or vice versa), the classification accuracy drops to less than 10 %. Having exhaustive training data to mitigate such domain shift issues shows to be crucial for the feasibility of the posture recognition system.

Chapter 9

Posture Recognition Demonstrator

Note: Parts of this chapter have been published by us in [89, 90]. This work hence exhibits similarities regarding formulations and visualizations.

By means of simulation, the previous chapter assessed the feasibility of a MI posture recognition system using purely passive tags. In this chapter we further extend this feasibility study by presenting an experimental demonstrator system. We repeat the single-frequency classification analyses from Cpt. 8, this time with experimental measurement data and verify its usability for posture recognition. Next, we introduce posture variations and coil displacements to our measurement data and find that these additional disturbances degrade the classification accuracy if they are not also present in the training. In order to improve robustness, we extend our system to use multi-frequency measurements of the input impedances and examine the resulting trade-off between bandwidth, center-frequency and classification accuracy.

9.1 Demonstrator and Measurement Setup

This section will describe our implementation of an MI posture recognition demonstrator in detail. The demonstrator's purpose is to verify the posture recognition capabilities that we stipulated in Cpt. 8 and to identify possible bottlenecks and other issues. It thus only represents the single next step towards a possible product by means of a proof of concept, but is in no way to be understood as a functioning prototype for end consumers. Some of the differences between this realization and our envisioned final system are summarized in Sec. 9.1.1. Directly after this juxtaposition we will introduce the electronics used for this demonstrator and explicate the considered postures and testing environments.

9.1.1 Demonstrator Structure

For the posture recognition demonstrator, we use the original solenoid proposal of Fig. 8.5a, which comprises 9 purely passive tag coils that are placed on the limbs and 4 measuring anchor coils that are situated on the torso. Each of these coils is fastened

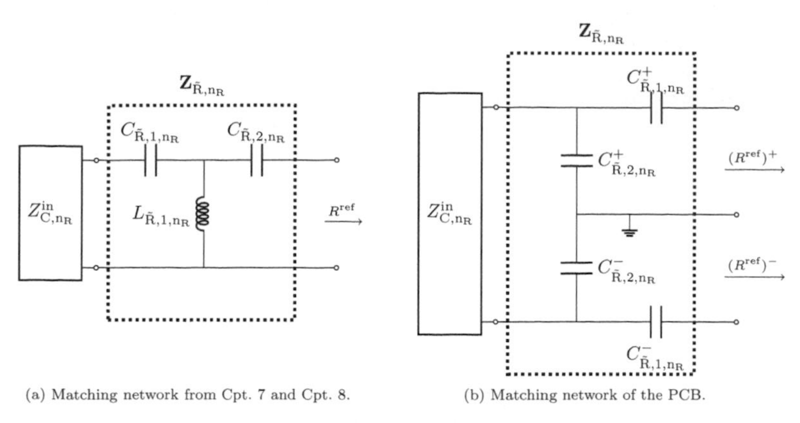

(a) Matching network from Cpt. 7 and Cpt. 8. (b) Matching network of the PCB.

Figure 9.1: Overview of different matching networks.

onto individual 3D-printed frames that can be worn on top of conventional clothing and also offer space for additional circuitry. The coils themselves are thin-wired using high-frequency litz wires. Despite operating in the kilohertz regime, these specific wires help to reduce the resistance and hence drastically increase the coils' quality factors and their coupling capabilities. In Tab. 9.1 we summarize the measured parameters of all coils without additional circuitry. For ease of use, both the hip and the torso coils have an ellipse shaped base, with the minor axis length being 0.85 times the major axis length. For these coils, the major axis length is stated as the diameter in the table. The passive tag coils are loaded with tuneable capacitors, which can be adjusted to obtain resonance at $f = 500\,\text{kHz}$ if no other coils are present. In contrast, the anchors are connected to a **P**rinted **C**ircuit **B**oard (PCB), which comprises matching networks, switches and other electronic components. The PCB is used to match each anchor, to switch it on and off via a single central unit, and to connect it to the VNA. The balanced connections from the PCB to the VNA are established via **S**ub**M**iniature version **A** (SMA) microwave cables with a $50\,\Omega$ resistance and all anchor impedances are measured sequentially. In Fig. 9.1b a PCB matching network is shown, which compared to Fig. 9.1a, i.e. the T-structured matching approach from Cpt. 7 and Cpt. 8, has two output ports to allow for balanced measurements. The measured data is saved and processed via connected computer. The system and measurement room are shown in Fig. 9.2.

In comparison, a final consumer system would likely use sewn-in coils which are already being used for smart garments [11]. Moreover, the system would likely deploy

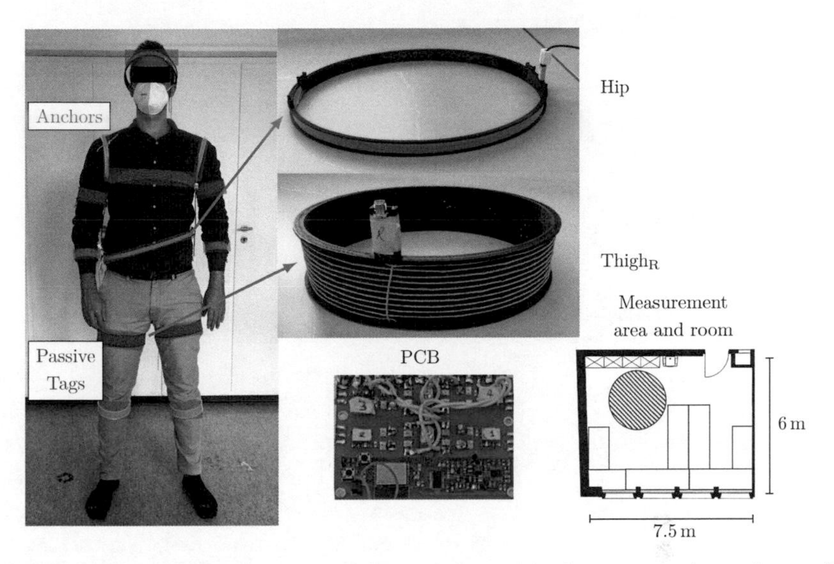

Figure 9.2: Photo of a test subject wearing the posture recognition demonstrator, close-up images of two of its coils, the PCB and the measurement room layout. The anchors (blue) and the tags (red) are mounted on wearable 3D-printed frames.

Parameter → Coil ↓		D^{coil} [cm]	N^{coil} [1]	H^{coil} [cm]	R [Ω]	L [µH]	C [pF]	f^{self} [MHz]	Q [1]
Receiving anchors	n_{R}								
$\text{Shoulder}_{\text{L}}$	1	17	5	1	0.1	10.1	49.9	7.1	244.1
$\text{Shoulder}_{\text{R}}$	2	17	5	1	0.2	10	48.3	7.2	174.5
Torso	3	37	10	4	0.5	56.2	83.2	2.4	353.1
Hip	4	39	10	1.5	0.2	26.7	209.3	2.2	419.4
Passive tags	n_{T}								
Calf_{L}	1	14	15	3	0.4	48.6	8.1	8	381.7
Calf_{R}	2	14	15	3	0.3	48.6	5.2	10	508.9
Thigh_{L}	3	20	10	4	0.3	32.6	0.4	46.1	341.4
Thigh_{R}	4	20	10	4	0.3	32.7	6.4	11	342.4
$\text{Forearm}_{\text{L}}$	5	9	15	3	0.2	26	5.3e-4	1.4e3	408.4
$\text{Forearm}_{\text{R}}$	6	9	15	3	0.2	25.9	1.1e-3	9.2e2	406.8
$\text{Upper arm}_{\text{L}}$	7	12	10	4	0.1	15.9	2.6e-4	2.5e3	499.5
$\text{Upper arm}_{\text{R}}$	8	12	10	4	0.1	15.9	1.2e-4	3.7e3	499.5
Head	9	21	10	2	0.5	42.9	8.9	8.1	269.5

Table 9.1: All coils and their measured parameters. All numerical values are rounded to single decimals, if applicable.

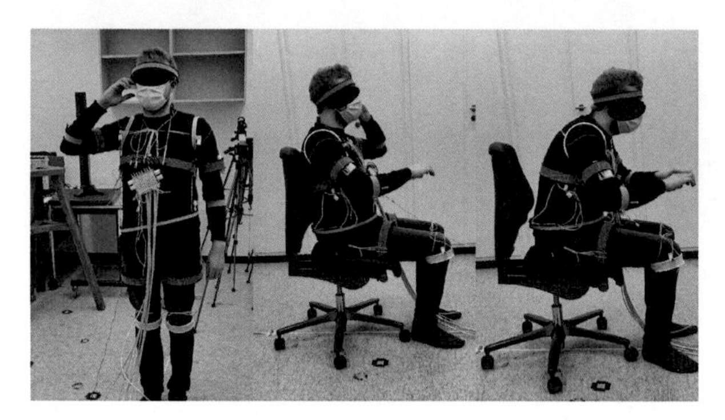

Figure 9.3: Sample postures.

an integrated circuit board which would already comprise a dedicated measurement circuit (e.g. similar to [138]). Such an integrated circuit board would simplify the measurement process, as it would not necessarily require matching networks and other electronic components of our PCB. Lastly, a consumer system may forward the measurement data via wireless communication standard such as Bluetooth or WLAN to a locally available mobile **C**entral **P**rocessing **U**nits (CPUs), e.g. via smartphone, where the entire signal processing (filtering, calibration and classification) takes place.

9.1.2 Data Collection and Pre-Processing

Overall, we consider the same postures as in Cpt. 8, which are sketched in Fig. 8.2. Some of the postures are also re-enacted by a test subject in Fig. 9.3. All subsequent measurements were taken in a laboratory that resembles an office environment (cf. Fig. 9.2). It contains desks, chairs, computers, monitors, drawers, various electronic components and measurement devices.

While conducting a measurement campaign, the fully equipped test subject took on each posture for as long as the measurement device needed to perform $N = 128$ frequency sweeps of the complex impedances. The frequency sweep used frequencies $f \in [480, 560]$ kHz at each anchor and with a spacing of 500 Hz totaling 161 frequency points. All anchor coils except the one being measured were open-circuited to limit the mutual impedance detuning of anchors. The switching between the anchor coils was performed automatically. The complex impedances were again partitioned into real and imaginary parts to obtain a real eight-dimensional feature vector (cf. (7.3)). In total, we thus obtained $128 \times 8 \times 161$ datapoints for each posture. Overall, we performed three such measurement campaigns in the same office environment, leading to three distinct datasets of all postures. We will refer to them as datasets 1, 2, and 3 throughout our work. Within each dataset, there are measurement variations due to the noise of the measurement device, due to the test subject not being able to stand completely still, and due to other external factors. For measurements belonging to different datasets, there are additional variations as a result of a slightly different placement of the coils due to removing and remounting them between measurement campaigns and also due to intentional minor changes of all postures. These inter-dataset variations are deliberate to emulate disturbances the system may be confronted with in everyday use.

In general, we further split the obtained measurement samples of each dataset into an individual training part containing 2/3 of each full dataset and a testing part containing the remaining other 1/3. This split is performed randomly over the $N = 128$ realizations, so all postures and frequencies are still equiprobable for both training and testing data.

9.2 Single-Frequency System

In this section, we only rely on the impedance measurements of dataset 1 at a frequency of $f = 500\,\text{kHz}$. While this choice reduces the overall available information and hence most likely the obtainable classification accuracy, it offers some practical advantages:

1. Only using a single frequency with a narrow bandwidth drastically reduces the hardware complexity of the involved circuitry as well as the measurement device.
2. The measurement speed is drastically increased and there is less data that needs to be transmitted and used for training.
3. The applicable classification algorithms can be simpler as we operate in a low, eight-dimensional feature space and hence do not have to consider advanced neural networks such as **C**onvolutional **N**eural **N**etworks (CNNs) [145] or **L**ong **S**hort-**T**erm **M**emory (LSTM) **R**ecurrent **N**eural **N**etworks (RNNs) [146], which are beneficial when dealing with highly-dimensional and possibly correlated sequences of features.

9.2.1 Single-Frequency Measurements

Before we analyze the classification performance, we first have a closer look at the impedance measurements of the full demonstrator. To this end, Fig. 9.4a and Fig. 9.4b show all $N = 128$ samples of dataset 1 for each anchor and posture in the complex plane. Each marker type represents a different anchor and each color represents a different posture. Note that due to the different structure of the balanced matching networks, additional detuning causes the anchor input resistances to be greater instead of lower than $50\,\Omega$. This effect is further enhanced by the lower coil resistances (cf. Tab. 9.1), which require a higher matching factor. For a given anchor and posture, we also observe impedance measurement deviations larger than $5\,\%$ of the corresponding mean value. Moreover, the distribution of these deviations is not symmetric as one would expect for additive Gaussian noise. Due to the pronounced asymmetry, we assume that the deviations are dominated by the test subject's involuntary movements during the measurements procedure.

Next, it is evident that the hip anchor (marker +) generally exhibits the strongest impedance detuning from the original $50\,\Omega$ and that this single anchor may distinguish many of the postures on its own, such as *Stand I* (pink) and *Stand Walk$_R$* (black). However, only using the hip anchor would be problematic when we need to differentiate between sitting postures, such as *Sit Call$_R$* (vermillion) and *Sit Front* (yellow). These postures are however easily distinguishable for the other anchors, which confirms that it is advantageous to use more than one anchor coil. When combining the observations of all anchors, i.e. using the full eight-dimensional feature vectors, the posture-based clustering of the impedance measurements becomes even more evident. This effect is illustrated by the **t**-distributed **S**tochastic **N**eighbor **E**mbedding (tSNE) plot [147] in Fig. 9.4c. While its results are not representative to quantitatively evaluate the true distances between the clusters, it shows that first, each eight-dimensional measurement sample of a given posture clusters well within a limited domain and second, the domains of each cluster are distinct. We thus expect that for the given dataset perfect accuracy can be achieved. Also, note that we consider the impedances differences for the left and right shoulder coils to be the result of non-identical coils and slightly asymmetric realizations of the postures.

9.2.2 Single-Frequency Classification Performance

To test this assumption of perfect accuracy and further analyze the classification performance, we train one-vs-one SVMs on the training data via five-fold cross validation using the same hyperparameters as in in Cpt. 8. Again, we also partition the impedance vector $\mathbf{x}_p$ that contains all four complex anchor input impedance measurements into a real eight-dimensional feature vector $\mathbf{r}_p$. As before, we choose the classification accuracy of the SVM on the testing set as our performance metric. With this configuration, the SVM yields a perfect accuracy of 1 and is hence able to correctly classify all postures (not shown). We reran this analysis for all three datasets and found that the perfect accuracy is always obtained.

SVM Performance for Additive Gaussian Impedance Noise So far, we assumed that the noise of the measurement device itself can be fully neglected compared to the impact of minor posture variations. However, for small low-cost devices or other external impacts this may generally not be true and we hence analyze the system's robustness against additional perturbations. To this end, we add circularly-symmetric complex white Gaussian noise with zero mean and standard deviation σ^{AWGN} onto the measured complex impedances to emulate a worsening noise characteristic.

In Fig. 9.5, we conducted a performance investigation of the SVM for this noisy dataset. We see that the SVM maintains an almost perfect accuracy until the noise standard deviation passes $\sigma^{\text{AWGN}} = 40\,\Omega$. For stronger noise levels we see a steady decline until $\sigma^{\text{AWGN}} = 10^4\,\Omega$, at which point the SVM has a classification accuracy of only $\frac{1}{14} \approx 7\,\%$, which equals a random guess. The 90 % accuracy is reached at a noise level of about $\sigma^{\text{AWGN}} = 100\,\Omega$, which makes up roughly 3 % of the maximally measured impedance.

ML Classifier for Additive Gaussian Impedance Noise Next, we want to provide the ML classifier for the case that each noisy sample is the result of (i) randomly selecting one of the $N = 128$ corresponding recorded impedance measurements with equiprobability and (ii) adding the Gaussian noise to emulate a worsening noise characteristic. The underlying assumption of this ML classifier is hence that all no values other than the $N = 128$ recorded impedance measurements per posture are possible and that the Gaussian noise sufficiently represents any further perturbation. The conditional PDFs of the corresponding noisy feature vectors $\mathbf{r}^{\text{noisy}}$ follow via Gaussian Mixture Models (GMMs) [148] with $N = 128$ equiprobable Gaussian components. More specifically, for a given posture p the random $\mathbf{r}^{\text{noisy}}$ are distributed according to

$$f_{\mathbf{r}^{\text{noisy}}}\left(\mathbf{r}^{\text{noisy}}|p\right) = \frac{1}{N}\sum_{n=1}^{N=128} f^{\text{AWGN}}\left(\mathbf{r}^{\text{noisy}};\, \mathbf{r}_{p,n},\, \frac{(\sigma^{\text{AWGN}})^2}{2}\mathbf{I}_8\right), \tag{9.1}$$

where $f^{\text{AWGN}}(.)$ is a real multivariate Gaussian PDF, $\frac{(\sigma^{\text{AWGN}})^2}{2}\mathbf{I}_8$ is the covariance matrix and the respective means are given by measured feature vectors $\mathbf{r}_{p,n}$ (without additional noise).

The ML hypothesis, which also coincides with the **M**aximum **A** **P**osteriori (MAP) hypothesis due to the equiprobable posture priors, is thus the posture which maximizes (9.1) for a given realization $\mathbf{r}^{\text{noisy}}$, i.e.

$$\hat{p} = \arg\max_{p} f_{\mathbf{r}^{\text{noisy}}}\left(\mathbf{r}^{\text{noisy}}|p\right) . \tag{9.2}$$

Yet, the evaluation of (9.1) requires to compute a sum of exponentials for a wide range of values and can lead to numerical complications. To bypass these complications, we can alternatively break down (9.2) to a series of pairwise decisions. That is, for each possible pair of postures p and q with $q \neq p$ we instead evaluate the likelihood ratio $\frac{f_{\mathbf{r}^{\text{noisy}}}\left(\mathbf{r}^{\text{noisy}}|p\right)}{f_{\mathbf{r}^{\text{noisy}}}\left(\mathbf{r}^{\text{noisy}}|q\right)}$ individually. When evaluating these likelihood ratios, we can offset the exponential sums of both conditional PDFs via multiplication by a fixed constant $\exp(\alpha_{p,q}^2)$ with $\alpha_{p,q} = \frac{\min_n\{\ \{\|\mathbf{r}^{\text{noisy}}-\mathbf{r}_{p,n}\|\},\{\|\mathbf{r}^{\text{noisy}}-\mathbf{r}_{q,n}\|\}\}}{\sigma^{\text{AWGN}}}$. Combining this offset with the pairwise evaluation can make the problem numerically feasible without loss of generality.

Lastly, in case the dataset is split into a training and testing part, this additional information can be incorporated into the ML classifier. To this end, instead of using all N real feature vectors $\mathbf{r}_{p,n}$ in (9.1), only the ones which correspond to the testing set are considered. For our investigation with fixed measurements and additional Gaussian noise, the ML classifier of (9.2) constitutes an statistical upper bound to any other possible classifier such as a SVM. However, it is unfeasible in practice as the feature vectors $\mathbf{r}_{p,n}$ of the testing data are typically unknown beforehand. Furthermore, the conditional PDF of (9.1) is only the result of our assumptions and the adequate incorporation of real world perturbations (e.g. due to more posture variations, interference, or distortions) may lead to a fundamentally different PDF, which in turn would require a different ML classifier. Nevertheless, we can use (9.2) as a theoretical benchmark for practical classifiers. Further, the simplified evaluation of (9.2) is computationally cheap and facilitates the study of the posture recognition system.

ML Classifier Performance for Additive Gaussian Impedance Noise In Fig. 9.5, we compare the performance of the ML classifier as a theoretical upper bound to that of the SVM. Regarding the average classification accuracy, the ML classifier exhibits a similar behavior and outperforms the SVM by at most 6 %, despite using the testing data which offers superior prior information. Due to these minuscule differences, we presume that the training data reasonably represents the testing data and that the SVM approximates the optimal decision boundaries fairly well. Overall, we summarize that the full system exhibits a great performance and shows to be resistant against additive Gaussian impedance noise.

In order to get a more detailed picture of the underlying system-related classification challenges for individual postures, Fig. 9.6a and Fig. 9.6b show the confusion charts of the ML classifier when averaging the accuracy over many noise realizations with $\sigma^{\mathrm{AWGN}} = 100\,\Omega$ and $\sigma^{\mathrm{AWGN}} = 300\,\Omega$, respectively. The confusion chart for $\sigma^{\mathrm{AWGN}} = 100\,\Omega$ has an overall accuracy of 92 % and the errors are dominated by misclassifications between the *Sit Call*$_R$ and the *Sit Front* postures, as well as the *Stand Fall*$_L$ and *Stand Fall*$_R$ postures. For this noise level, there were also no misclassifications between the super classes *Sit* and *Stand*. For $\sigma^{\mathrm{AWGN}} = 300\,\Omega$, the overall accuracy is decreased to 67 % and now includes various different posture misclassifications, even between super classes. However, even for this high noise level some postures are recognized well, such as *Sit I*, *Stand Walk*$_R$ or *Stand X*. To understand why some of these misclassifications are dominant, we next consider all pairwise Euclidean inter-posture distances between all eight-dimensional feature vectors of one posture and all feature vectors of all other postures, in the absence of the additive noise. Fig. 9.7 illustrates these inter-posture distances via box plots, where the whiskers represent the true minimum and maximum distances, the box itself shows the interquartile range (25th to 75th percentile) and the red bar shows the median distance. As in Sec. 8.3, larger Euclidean distances for a given posture hence indicate a higher robustness against misclassifications. The box plot consequently shows that misclassifications between both *Stand Fall* postures or between the *Sit Call*$_R$ and the *Sit Front* postures are more likely to occur and hence reaffirms the previous results of the confusion charts. To mitigate these errors efficiently, it is necessary to adjust the system itself, e.g. by adapting the coil design or placement, such that the detuning of either of these postures increases.

Different Anchor Selections Based on the ML classifier Another interesting question is which of the anchors are actually needed when additional disturbances, as described in the previous paragraph, are present. Although it is clear that the hip (H) and torso (T) anchors are detuned the most, we saw that the inclusion of the shoulder anchors ($\mathrm{S_L}$ and $\mathrm{S_R}$) may be beneficial to distinguish specific postures. To quantify these considerations, Fig. 9.8 shows the average accuracy of the ML classifier for all possible anchor selections. For this extensive analysis, we used the ML classifier instead of the SVM to reduce computation times, but expect a similar overall behavior (cf. Fig. 9.5). The anchor selection is expressed in terms of an ordered binary sequence,

S_L S_R T H, where a 1 indicates that the anchor is used and a 0 represents an anchor unused. The sequence 0 1 1 0 for example means that only observations from the right shoulder and the torso anchor are used. The sequences are ordered from the worst to the best in terms of accuracy averaged over all noise levels. As a first observation, we see that when only using a single anchor, the left shoulder anchor performs worst, closely followed by the right shoulder anchor, then the torso anchor and lastly the hip anchor (rows $1, 2, 4$ and 8). While none of the anchors manages to obtain a perfect accuracy individually, a system with reduced complexity that e.g. only uses the left shoulder and hip anchor (row 10) may still be sufficient for some fields of application.

SVM Performance Degradation under a Domain Shift (Mismatch) Up to now, the experimental results hold qualitatively for any of the datasets 1, 2, or 3, and affirm a high classification accuracy. In a next step, we analyze the generalizability between those datasets. That is, Fig. 9.5 also shows the performance of a mismatched SVM which was trained on one of datasets and tested on the other two. This mismatched scenario shows a significant degradation compared to the correctly trained SVM and exhibits a classification accuracy of roughly 68 %, even for low noise levels. Surprisingly, the accuracy improves slightly for minor increases of the noise level, before it starts to decline and approach the accuracy of a random guess. As in Sec. 8.5.3, we expect the bad initial performance to be the result of shifted impedance clusters for some postures. For such postures, the decision boundaries learned from one dataset may not at all enclose the corresponding impedance cluster from a different dataset. As a result, misclassifications are present even if the additional noise is weak. Yet, for increasing noise levels the perturbations can (i) randomly compensate this shift or (ii) lead to broader decision boundaries, which may explain the slight improvement of the classification accuracy. Overall, these results highlight that domain shifts are a practical issues that significantly reduce the usability of the experimental system and need to be addressed.

9.3 Multi-Frequency System

The experimental results of Sec. 9.2.2 reinforce our simulation-based findings from Cpt. 8, which showed the presented low-complexity single-frequency system to work with high accuracy. This high accuracy is maintained even when only relying on two anchor coils (cf. Fig. 9.8). However, the simulation and experimental study both revealed that domain shifts, i.e. a mismatch between the PDFs of training and testing data, can cause a severe performance degradation (cf. Fig. 8.12 and Fig. 9.5). This performance degradation overshadowed the impact of additional noise significantly and rendered the posture recognition system unusable. In Cpt. 8, the mismatch was analyzed by using training data from one body model and testing data from a different body model, which may in practice be circumvented by requiring each individual user to calibrate their own system. Alternatively, it could also be mitigated by having exhaustive training data or by applying transformations that allow to match more general training data to a specific user. In contrast, a mismatch that is caused by irregular posture variations and minor coil displacements would be harder to avoid in everyday use. The experimental single-frequency study already revealed that such issues are also detrimental and need to be addressed. This section hence investigates this latter type of mismatch and further illustrates why single-frequency measurements are sometimes insufficient for reliable posture recognition. Moreover, we qualitatively analyze the advantages of using a broader measurement spectrum to deal with such domain shifts at the cost of a more complex system. In this process, we characterize the trade-off between bandwidth, center frequency and classification accuracy.

9.3.1 Multi-Frequency Measurements and Domain Shifts

Fig. 9.9 depicts multiple measured input impedances of the *hip* anchor for two different postures at a single frequency of $f = 495\,\mathrm{kHz}$. The postures are distinguished by color and the different datasets to which the measurements belong are distinguished by marker type. We observe that the impedance variations within each dataset are minor compared to the differences across the different datasets. We also see that the *Stand* $Call_L$ posture shows overall weaker variations compared to the *Stand* X posture, both within each dataset as well as across the different datasets. The figure also includes a decision boundary (dashed line), which only considers observations of dataset 1 as training data. This boundary clearly works perfectly for any impedance of either dataset 1 or dataset 2. For dataset 3 however, all impedances of the *Stand* X posture would incorrectly be classified as *Stand* $Call_L$ as a result of the previously described mismatch. In comparison, if impedances of all three datasets were considered, the resulting modified decision boundary represented by the solid line correctly distinguishes impedances from all datasets. We conclude that for single-frequency measurements, the distribution of one dataset does not represent the distributions of the other datasets sufficiently well. This result is common for supervised classifiers and illustrates the need for comprehensive training data or other adaption techniques [149]. Obtaining more comprehensive and diverse training data via extensive additional measurements is possible, but requires a substantially higher user effort. Alternatively, this domain shift issue may also be solved by using augmented or transferred training data which however requires a very accurate model and precise measurements of the coils and noise to work well. Lastly, it is also possible to incorporate different features (other than the single-frequency complex impedances) which carry more posture information or are more robust against the described mismatch. It is well known, that magnetic coupling affects the resonance frequency of the circuits involved. If the resonant circuits have a high quality factor, this detuning may result in considerable impedance changes at a given frequency. While this effect is desired in principle as it allows to detect posture-induced changes in the coil topology, it also may explain the observed inter-dataset variations that are exemplified in Fig. 9.9. We anticipate for these reasons, that using wideband impedance measurements as extended features may proof to be more robust. To investigate this conjecture, Fig. 9.10 shows the same resistances and reactances for all three datasets, but now measured over all frequencies. First, it can be observed that despite each passive tag being resonant at $f = 500\,\mathrm{kHz}$, the overall resonance peaks are shifted to higher frequencies as a result of the compound near-field coupling. This means, that depending on the posture, different frequencies may be more relevant to characterize the detuning behavior and no single frequency may reflect the true detuning behavior properly for all different postures. We also find that the shape of the impedance curves is distinct for different postures. However, the shape of the impedance curves does not vary significantly between either measurements or datasets. This similarity highlights that the differences observed in Fig. 9.9 are minuscule (Ω scale) compared to the overall detuning behavior over frequency (kΩ scale). We therefore conjecture that learning the impedance behavior over frequency instead of using a single-frequency measurement makes a classifier more robust when training data and testing data are taken from different datasets.

Fig. 9.11 shows the eight standardized feature sequences, i.e. real and imaginary parts of the four anchor impedances over frequency, as rows of an image. The standardization was performed for each feature sequence individually (e.g. the real part of the hip anchor impedance over frequency), and the corresponding single mean and single standard deviations for each sequence were empirically determined via training data (cf. Sec. 9.3.2). The anchor names are again abbreviated as $\mathrm{S_L}$ (shoulder left), $\mathrm{S_R}$ (shoulder right), T (torso), and H (hip). The fourth and eighth row of Fig. 9.11 are for example scaled and shifted versions of one corresponding line (marker x) from

Fig. 9.10. Ultimately, such a 8×161-dimensional image is a simple approach to compactly visualize one multi-anchor multi-frequency measurement, which is used as input data for the classification process. For the two postures shown in Fig. 9.11, it is for example easily observed that the left shoulder anchor $\mathrm{S_L}$ is more helpful to distinguish between the two postures than its right counterpart $\mathrm{S_R}$.

Overall, we conclude that an extended frequency spectrum is helpful to observe the full extent of the posture-induced impedance detuning. At least at a first glance, the multi-frequency measurement also seem to be more robust and may hence mitigate the performance degradation caused by domain shifts. However, using multiple frequencies requires more intricate measurement circuits and more measurement time. Moreover, since the amount of data that needs to be processed increases with this approach, it also leads to longer computation times.

9.3.2 Multi-Frequency Classification Performance

As described in Sec. 9.1.2, each of the datasets 1, 2 and 3 contains $N = 128$ of the 8×161-dimensional images per posture (cf. Fig. 9.11), of which $\frac{2}{3}$ are used for the training data and $\frac{1}{3}$ for the testing data.

For the multi-frequency classification, we compare two widely-known supervised classification algorithms: a SVM, which was already used for the single-frequency measurements (cf. Sec. 9.2.2), and a CNN. The latter choice is more complex and was unnecessary for the low-dimensional feature space of the single-frequency operation. In contrast, it may be beneficial for the multi-frequency operation as it is capable to factor in important correlations inbetween and within different feature sequences.

1. The input for the SVM is a flattened (vectorized) $1 \times 8N_F$-dimensional vector of the corresponding $8 \times N_F$-dimensional image, where N_F is the number of considered frequencies from our measurements with $1 \leq N_F \leq 161$. The SVM uses a RBF kernel and the other hyperparameters are determined via random search.

2. The CNN directly uses the $8 \times N_F$-dimensional image as standardized input with three convolutional layers containing $12, 24$ and 48 filters with a kernel size of 2 and a stride of 1. Each of these layers is followed by a batch normalization layer, a rectified linear unit activation and a max pooling layer. After these convolutions, a fully connected layer is directly followed by a softmax normalization. Subsequently, the posture which corresponds to the maximum output is selected.

The classifiers' accuracy on the testing data is compared in Fig. 9.12. For the case *General* (opaque bars), the classifiers are trained on the training data of all three datasets and their accuracy is examined on the corresponding testing data of all three datasets (comparable to the solid decision boundary of Fig. 9.9). For the case *Mismatch* (transparent bars), each classifier is only trained on the training data of one single dataset and tested on the testing data of the other two datasets (comparable to the dashed decision boundary of Fig. 9.9). This is done for all datasets 1,2 and 3 with the accuracy being averaged. The figure further distinguishes between using impedances from all frequencies (f_{all}, red) with $N_F = 161$ and only using the impedances at $f = 500\,\text{kHz}$ (f_{500}, blue) with $N_F = 1$.

We observe that in all *General* scenarios, the classification is perfect, regardless of the used classifier or the used frequency range. As we now do not consider additional noise, this observation is in line with our findings from Sec. 9.2.2. In contrast, in the scenario *Mismatch* with single-frequency observations, the accuracy degrades to 65 % for the SVM and to 56 % for the CNN. This performance loss likely is the result of the decision boundaries not being general enough to deal with the variety of unknown new data, as was stipulated in Sec. 9.3.1. By extending the measurement spectrum, this loss of accuracy can be mitigated to 75 % for the SVM and 91 % for the CNN.

While the multi-frequency classifiers do not yield the same results as using more comprehensive training data, they still lead to a reasonable robustness when being tested on new data with additional disturbances. This approach hence seems a viable trade-off in case comprehensive training data cannot be acquired.

9.3.3 Complexity Reduction via Feature Simplification

After quantifying the possible robustness gain of using multi-frequency classifiers, we provide a more in-depth look into which features are truly necessary to obtain this performance gain. To do so, we reduce the amount of used features in a practical manner, such that is it mitigates the inherent complexity increase of the multi-frequency operation. That is, we repeat the full classification process but only selected a limited subset of all available features. We do not perform a feature transformation or extraction in a traditional machine learning manner, e.g. via principal component analysis or by analyzing the trained networks in detail, since the results may still require all features but weighted or combined differently. Such traditional approaches hence lead to a reduction of the computational complexity but might not necessarily affect the hardware complexity as well.

Measurement Spectrum: While the expansion from a single-frequency scenario to a multi-frequency scenario led to noticeable robustness advantages, not all frequencies may be equally relevant. A reduction of unnecessary frequency sampling points would be beneficial as it would mitigate the complexity issues described in Sec. 9.3.1. In the following, we focus on the *Mismatch* scenario and use the same CNN architecture as before. Yet, we now train and test the CNN only on a reduced subset of all available frequencies. In Fig. 9.13 the results of this analysis are visualized as intensity plot, where the obtained CNN classification accuracy is shown for different combinations of center frequency and bandwidth. The plot shows that an increasing bandwidth generally improves the performance and that for a 90 % accuracy at least 10 kHz of bandwidth are required. As for the center frequency, the interval $[510, 530]$ kHz proves to be preferable with a minor performance advantage at the higher frequencies when using a low bandwidth. This is in line with the observations from Fig. 9.11, which showed this area to contain the most significant detuning. Moreover, the impedance measurements for frequencies $f \geq 540$ kHz are clearly unfavorable to distinguish between postures. Lastly, we also observe that the bandwidth can be halved or even quartered from 80 kHz to 40 kHz or 20 kHz without suffering a meaningful performance loss.

Magnitude of Impedance: Lastly, we want to analyze whether the simultaneous use of resistances and reactances is actually beneficial for the classification. Depending on the implementation of the impedance measurement device, it can be more convenient to instead rely on the magnitude of the impedance, which can be determined with non-coherent measurement units. Fig. 9.14 thus shows the resulting performance for such a change. The figure is analogous to Fig. 9.12 but instead of using all frequencies, it additionally illustrates the case f_{selected}, which uses a bandwidth of 40 kHz (i.e. $N_F = 81$) with a center frequency of 510 kHz. That is, the classifiers are now trained and tested using either 4×81-dimensional or 4×1-dimensional images[1] of the impedance magnitude measurements.

The performance of any of the shown scenarios is similar to our previous results on complex impedances, proving that neither the reduction of the used frequencies nor the switch from coherent to non-coherent impedance measurements causes any relevant performance loss.

[1] In case of the SVM, the flattened equivalent is used.

9.4 Conclusions

In this chapter, we built a posture recognition system based on the concept of Cpt. 7 and investigated it experimentally. A single-frequency study was conducted and showed that the system classified all postures correctly when a high-end measurement device was used. Further, the system still maintained a 90 % accuracy if the measurements were perturbed by additional errors with a standard deviation of more than 3 % of the maximally measured impedance value. It was further demonstrated that the SVM as a practical classifier exhibited a similar performance to that of an unfeasible and purely theoretic ML classifier. Yet, domain shifts as a result of intentional posture variations and minor coil displacements between datasets showed to degrade the system substantially. We analyzed the issues caused by these domain shifts in detail and consequently extended the system to use multi-frequency impedance measurements for an increased robustness. In this process, we characterized the trade-off between operating frequency, bandwidth and classification accuracy quantitatively. Lastly, it was shown that using measurements of the impedance magnitudes with a bandwidth of $B = 40\,\mathrm{kHz}$ and a operating frequency of $B = 510\,\mathrm{kHz}$ was sufficient to be robust against the domain shifts and obtain a classification accuracy of 89 %. This performance was obtained in conjunction with a CNN, which showed to be superior for multi-frequency impedance measurements.

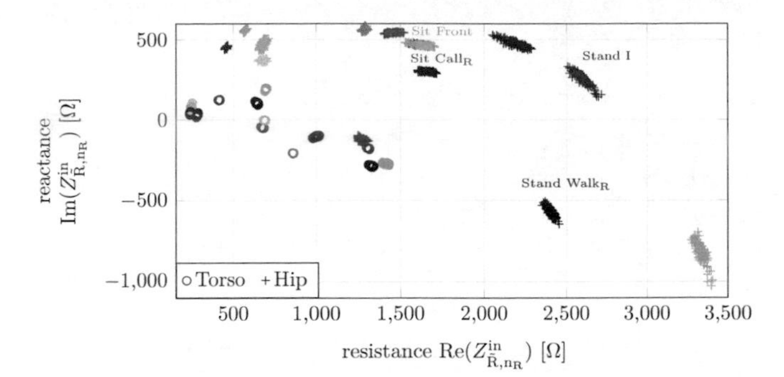

(a) Hip and torso anchor impedances.

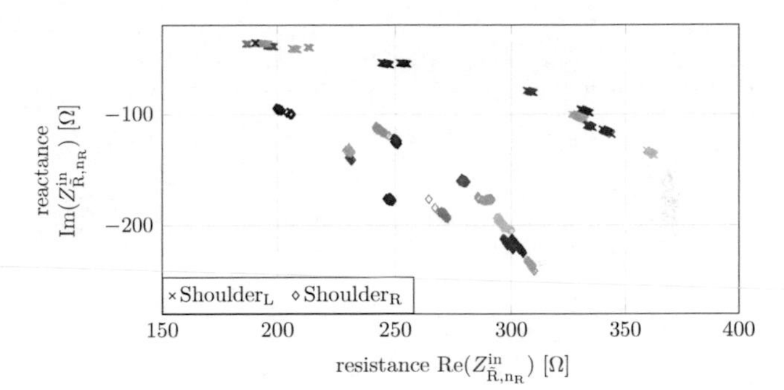

(b) Shoulder anchor impedances.

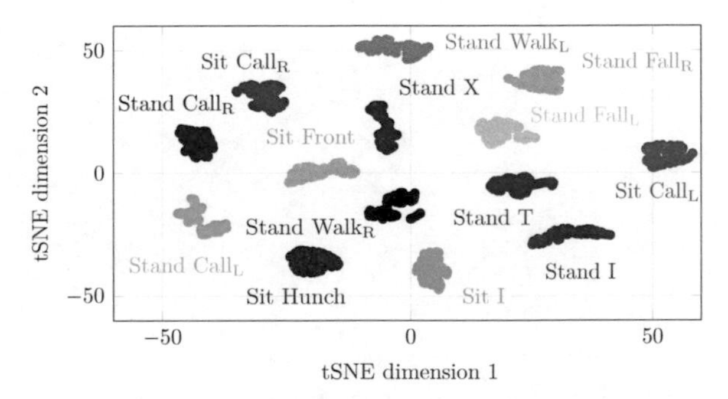

(c) tSNE plot when combining all anchors.

Figure 9.4: Various illustrations of the impedance measurements for all anchors at $f = 500\,\text{kHz}$. (a) and (b) show the measured input impedances of each anchor and for each posture in the complex plane. (c) shows the corresponding approximate clustering via tSNE plot when combining all four complex anchor impedances.

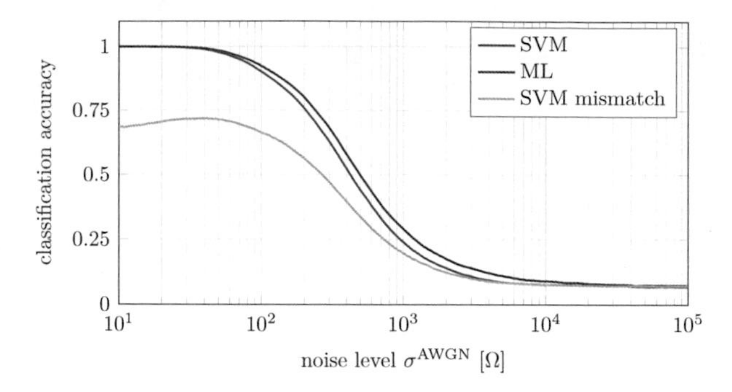

Figure 9.5: Comparison of the classification accuracy if the impedance measurements are subject to additional circularly-symmetric complex Gaussian noise with zero mean and varying standard deviations σ^{AWGN}.

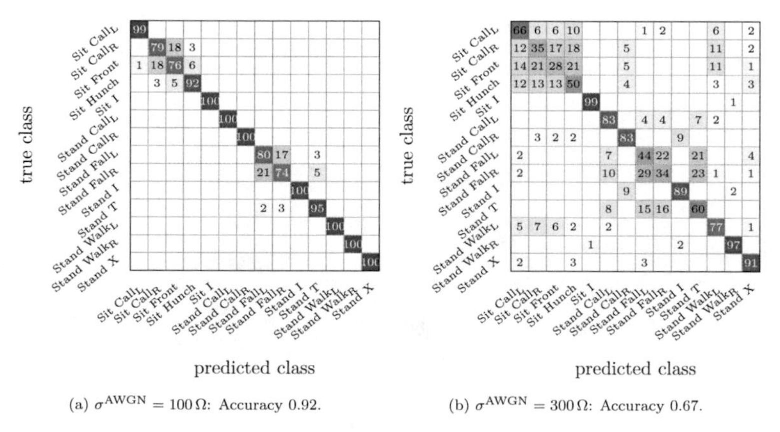

(a) $\sigma^{AWGN} = 100\,\Omega$: Accuracy 0.92.

(b) $\sigma^{AWGN} = 300\,\Omega$: Accuracy 0.67.

Figure 9.6: Confusion charts of the ML classifier displaying the average percentage of classifications for a low and a high noise level causing different numbers and types of misclassifications. All entries are rounded to integers, so the sum of each row or column might be unequal to 100.

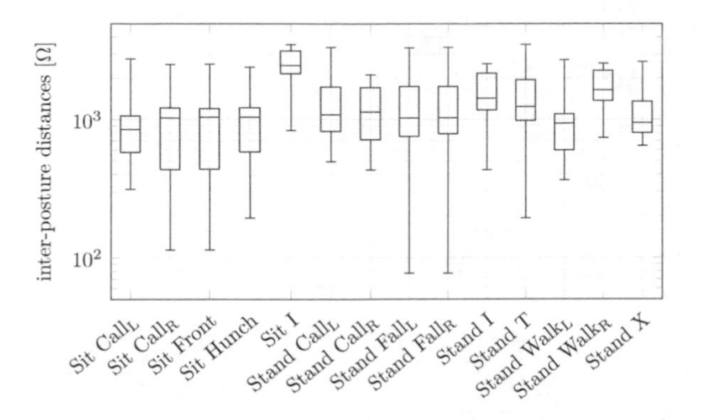

Figure 9.7: Box plot of all pairwise Euclidean inter-posture distances for all feature vectors without additional noise.

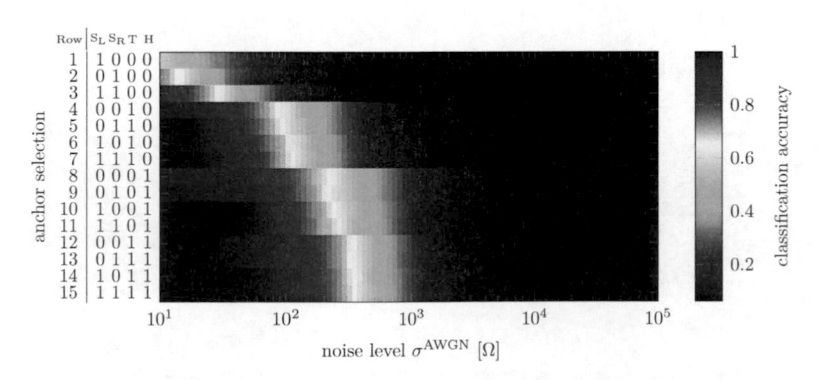

Figure 9.8: Overview of the accuracy that is obtained when using the ML classification for different choices of the four anchors: shoulder left (S_L), shoulder right (S_R), torso (T) and hip (H). The topology choices are ordered from worst (top row) to best (bottom row) with respect to their their average accuracy over all noise levels.

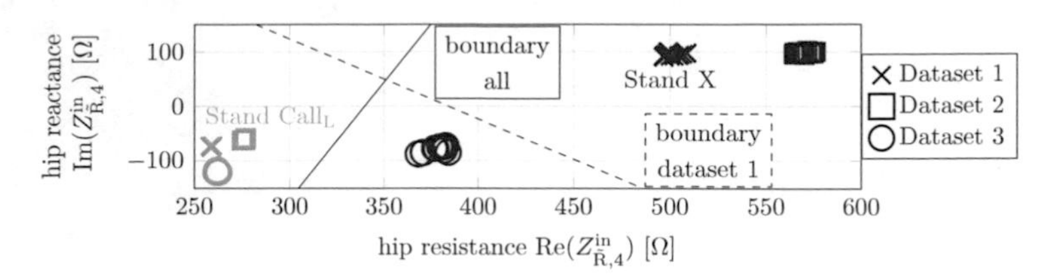

Figure 9.9: Single-frequency impedances of the Hip anchor for the *Sit Call$_L$* and *Stand X* postures at $f = 495\,\mathrm{kHz}$. The measurements are taken from three different datasets, which are subject to both posture variations and coil displacements.

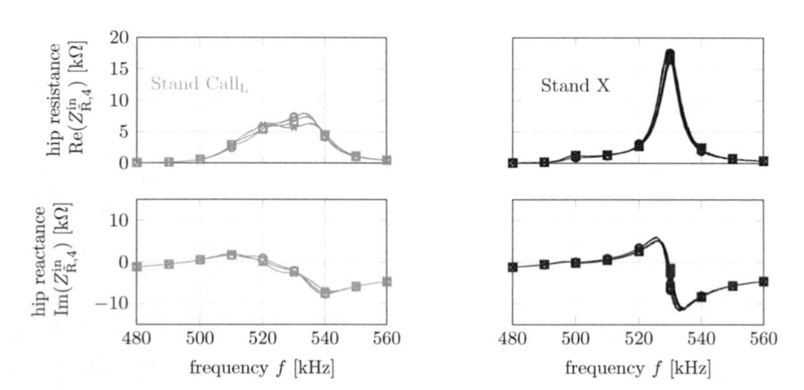

Figure 9.10: Multi-frequency impedances of the Hip anchor for the *Sit Call$_L$* and *Stand X* postures for the same different datasets.

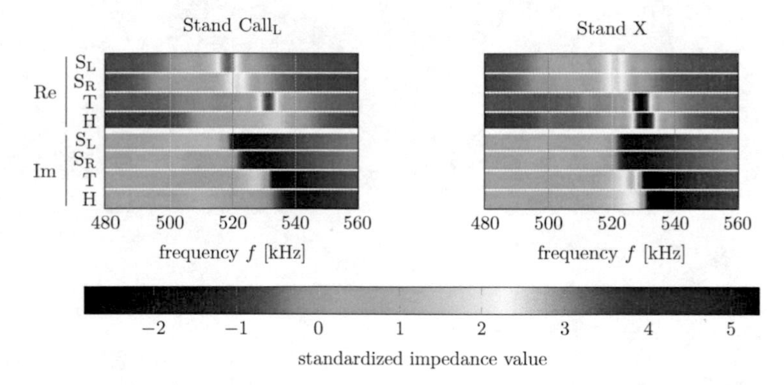

Figure 9.11: Full 8 × 161-dimensional images from dataset 1 for the *Sit Call$_L$* and *Stand X* postures. Each image comprises all eight standardized feature sequences for the frequencies $f \in [480, 560]\,\mathrm{kHz}$ with a 500 Hz spacing.

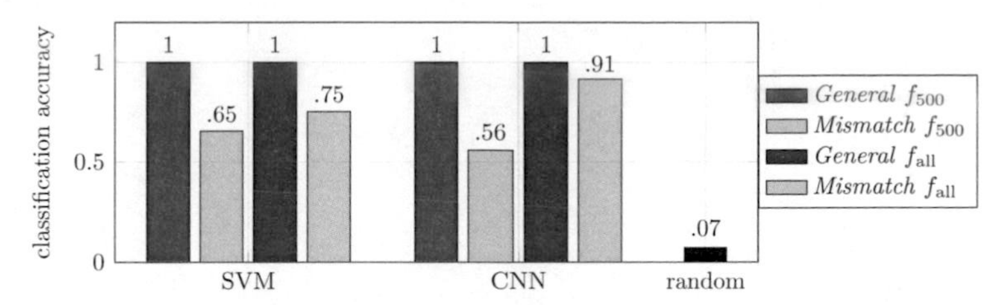

Figure 9.12: Comparison of different classifiers using either a single-frequency measurements (f_{500}) or the full bandwidth of 80 kHz (f_{all}). The case *General* uses samples of all three datasets for training and testing, while *Mismatch* only trains on one dataset and tests on the other two datasets.

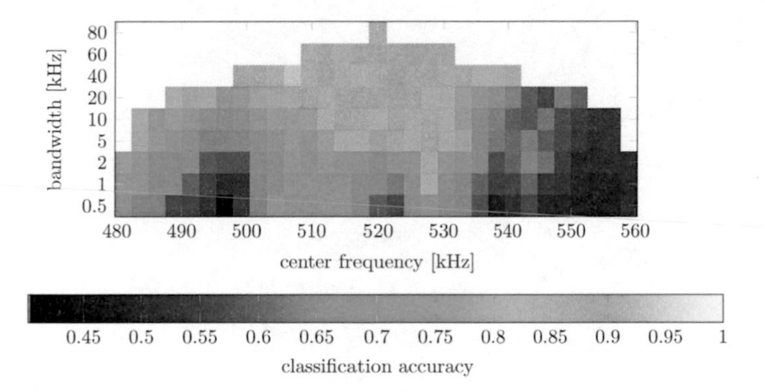

Figure 9.13: Impact of different measurement bandwidths and center frequencies on the classification accuracy for CNNs when analyzing the scenario *Mismatch*, i.e. training and testing is performed on strictly different datasets.

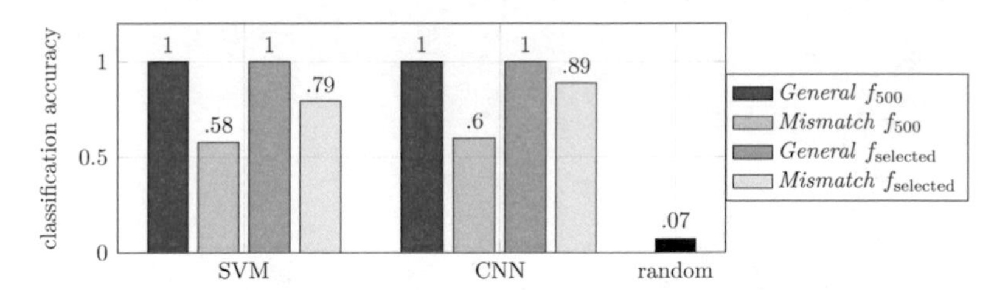

Figure 9.14: Comparison of different classifiers using either the single-frequency measurements (f_{500}) or a bandwidth of 40 kHz centered at 510 kHz ($f_{selected}$). The classifiers only used the magnitude of the impedances, compared to the analogous plot of Fig. 9.12, which used the complex impedances.

Chapter 10

Summary

This thesis investigated the use of low-frequency magnetic induction for body-centric tasks, such as localization and posture recognition. The following chapter summarizes our core results.

We extended a multiport system model to represent the MI near-field coupling of arbitrarily arranged coil antennas, which could be operated either actively or passively. We derived the associated PEB for this system model, which is the CRLB on the position RMSE and as such represents the maximum achievable localization accuracy of any unbiased estimator. The PEB revealed a drastically varying localization behavior depending on whether the system operates in a weakly-coupled or a strongly-coupled regime. For the weakly-coupled regime, the localization of passive agents showed to be unfeasible. For active agents in this regime, we characterized the scaling behavior with respect to common design parameters. We further illustrated that based on the network constellation, the measured signal exhibits a lack of spatial information about the agent in only a few limited directions. Estimation errors in these directions hence dominate the overall RMSE and degrade it by orders of magnitude, even for high SNRs. This lack of information in certain directions cannot easily be mitigated by using larger coil dimensions or other design parameters and may be particularly pronounced if the deployment area is unknown or if the placement area of the measuring anchors is restricted. While this directionality issue also occurs in the electromagnetic far field, it is more distinct in the near field due to its inherent sextic pathloss.

We showed that this issue can be managed if auxiliary coils, so called passive relays, are introduced to the network. These passive relays provide an additional signal contribution which fully counteracts the prior directionality issues if they are placed beneficially. Yet, if they couple too strongly with the agents, they further degrade the localization accuracy. We hence proposed low- and high-complexity switching schemes of the relay loads and investigated them for random and possibly dense network constellations. These switching schemes are not only capable of fully compensating any adverse effects of the relays, they also boost the median accuracy by almost two orders of magnitude and enhance the overall reliability of the localization.

We further presented the first study of cooperative localization for MI networks by allowing the exchange of channel state information between all coils. This approach shows to be an alternative means to mitigate the directionality issues and increase the SNR at the cost of a higher system complexity. We quantified this trade-off for different coil designs and numerical estimation methods. For a single pair of three-axis coils, we additionally derived the closed-form ML position estimator as a non-cooperative low-complexity option for MI localization.

We proposed a novel approach for the classification of MI network topologies, which uses resonantly loaded purely passive coils distributed on the human body and active coils that measure their input impedances. The mere presence of the passive coils detunes the impedances of the active coils, and the extent of this impact changes for different network topologies. This relationship between topology and input impedance can be learned via supervised classifiers. This novel classification approach was further used to create and study a wearable MI posture recognition system. We demonstrated the feasibility of this system by simulation, considering different types of perturbations, body models and coil designs. Even for single-frequency measurements of the impedance, reasonable operating points allowed for a classification accuracy of more than 90 %. This performance was retained for any of the supervised classifiers considered.

With the insights of the simulation-based study, we built a demonstrator of this posture recognition system to verify our results experimentally. Using measurements recorded in an office environment, a perfect classification accuracy was obtained when relying on a high-end measurement device. In contrast, a 90 % classification accuracy could be obtained via supervised classifiers when additional measurement errors were present, whose standard deviation was larger than 3 % of the maximally measured impedance. We further compared these supervised classifiers to the herein stated theoretic ML classifier and found no meaningful difference. However, for severe mismatches between the training and testing data, e.g. as a result of posture variations or displacements of the coils, the classification accuracy degraded to less than 60 %. We further showed that multi-frequency impedance measurements are robust to such mismatches and in combination with a CNN retained the previous high accuracy. We further quantified this trade-off between bandwidth, center frequency and classification accuracy. Lastly, we evaluated the option of using simpler measurement variables such as the impedance magnitudes and found those to be an adequate low-complexity alternative.

Appendix A

System Model Derivatives and Fisher Information

In order to calculate the estimation bounds that were introduced in Sec. 3.4, we need to evaluate the derivatives of the all quantities of our system model. In this appendix, we will do so iteratively from the top to the bottom with respect to an arbitrary scalar derivation parameter $[\boldsymbol{\Psi}]_j$ of the full agent constellation $\boldsymbol{\Psi}$.

Starting with the derivatives of the input currents, we have

$$\begin{aligned}\frac{\partial \mathbf{i}_{\mathrm{T}}^{\mathrm{in}}}{\partial[\boldsymbol{\Psi}]_j} &= \left[\frac{\partial \mathbf{G}^{\mathrm{active}}}{\partial[\boldsymbol{\Psi}]_j} - \mathbf{G}^{\mathrm{active}}\,(R^{\mathrm{ref}}\mathbf{I}_{N_{\mathrm{T}}} + \mathbf{Z}_{\tilde{\mathrm{T}}}^{\mathrm{out}})^{-1}\frac{\partial \mathbf{Z}_{\tilde{\mathrm{T}}}^{\mathrm{out}}}{\partial[\boldsymbol{\Psi}]_j}\right] \\ &\quad \cdot (R^{\mathrm{ref}}\mathbf{I}_{N_{\mathrm{T}}} + \mathbf{Z}_{\tilde{\mathrm{T}}}^{\mathrm{out}})^{-1}\,R^{\mathrm{ref}}\,\mathbf{i}_{\mathrm{T}}\,, \end{aligned} \tag{A.1}$$

$$\frac{\partial \mathbf{i}_{\mathrm{R}}^{\mathrm{in}}}{\partial[\boldsymbol{\Psi}]_j} = (R^{\mathrm{ref}}\mathbf{I}_{N_{\mathrm{R}}} + \mathbf{Z}_{\tilde{\mathrm{R}}}^{\mathrm{in}})^{-1}\frac{\partial \mathbf{Z}_{\tilde{\mathrm{R}}}^{\mathrm{in}}}{\partial[\boldsymbol{\Psi}]_j}\left[\mathbf{I}_{N_{\mathrm{R}}} - (R^{\mathrm{ref}}\mathbf{I}_{N_{\mathrm{R}}} + \mathbf{Z}_{\tilde{\mathrm{R}}}^{\mathrm{in}})^{-1}\mathbf{Z}_{\tilde{\mathrm{R}}}^{\mathrm{in}}\right]\mathbf{i}_{\mathrm{R}}\,. \tag{A.2}$$

As we only consider fixed and non-adaptive matching networks, the current gain $\mathbf{G}_{\tilde{\mathrm{R}}:\mathrm{RC}}$ of the anchor side is deployment independent. However, the current gain from the source to the agents' coil antennas depends on their deployment due to a possible detuning of $\mathbf{Z}_{\tilde{\mathrm{T}}}^{\mathrm{out}}$. The current gain derivatives hence follow as

$$\frac{\partial \mathbf{G}^{\mathrm{active}}}{\partial[\boldsymbol{\Psi}]_j} = \mathbf{G}_{\tilde{\mathrm{R}}:\mathrm{RC}}\left[\frac{\partial \mathbf{G}_{\mathrm{C}:\tilde{\mathrm{R}}\tilde{\mathrm{T}}}}{\partial[\boldsymbol{\Psi}]_j}\,\mathbf{G}_{\tilde{\mathrm{T}}:\mathrm{CT}} + \mathbf{G}_{\mathrm{C}:\tilde{\mathrm{R}}\tilde{\mathrm{T}}}\,\frac{\partial \mathbf{G}_{\tilde{\mathrm{T}}:\mathrm{CT}}}{\partial[\boldsymbol{\Psi}]_j}\right], \tag{A.3}$$

$$\frac{\partial \mathbf{G}_{\tilde{\mathrm{T}}:\mathrm{CT}}}{\partial[\boldsymbol{\Psi}]_j} = -(\mathbf{Z}_{\mathrm{C}}^{\mathrm{out}} + \mathbf{Z}_{\tilde{\mathrm{T}}:\mathrm{C}})^{-1}\frac{\partial \mathbf{Z}_{\mathrm{C}}^{\mathrm{out}}}{\partial[\boldsymbol{\Psi}]_j}(\mathbf{Z}_{\mathrm{C}}^{\mathrm{out}} + \mathbf{Z}_{\tilde{\mathrm{T}}:\mathrm{C}})^{-1}\mathbf{Z}_{\tilde{\mathrm{T}}:\mathrm{CT}}\,, \tag{A.4}$$

$$\frac{\partial \mathbf{G}_{\mathrm{C}:\tilde{\mathrm{R}}\tilde{\mathrm{T}}}}{\partial[\boldsymbol{\Psi}]_j} = (\mathbf{Z}_{\tilde{\mathrm{R}}}^{\mathrm{out}} + \mathbf{Z}_{\mathrm{C}:\tilde{\mathrm{R}}})^{-1}\left[\frac{\partial \mathbf{Z}_{\mathrm{C}:\tilde{\mathrm{R}}}}{\partial[\boldsymbol{\Psi}]_j}(\mathbf{Z}_{\tilde{\mathrm{R}}}^{\mathrm{out}} + \mathbf{Z}_{\mathrm{C}:\tilde{\mathrm{R}}})^{-1}\mathbf{Z}_{\mathrm{C}:\tilde{\mathrm{R}}\tilde{\mathrm{T}}} + \frac{\partial \mathbf{Z}_{\mathrm{C}:\tilde{\mathrm{R}}\tilde{\mathrm{T}}}}{\partial[\boldsymbol{\Psi}]_j}\right]. \tag{A.5}$$

The remaining derivatives of the input and output matrices can be calculated analogously as

$$\frac{\partial \mathbf{Z}_{\tilde{\mathrm{T}}}^{\mathrm{out}}}{\partial[\boldsymbol{\Psi}]_j} = \mathbf{Z}_{\tilde{\mathrm{T}}:\mathrm{CT}}^{\mathrm{T}}(\mathbf{Z}_{\tilde{\mathrm{T}}:\mathrm{C}} + \mathbf{Z}_{\mathrm{C}}^{\mathrm{out}})^{-1}\frac{\partial \mathbf{Z}_{\mathrm{C}}^{\mathrm{out}}}{\partial[\boldsymbol{\Psi}]_j}(\mathbf{Z}_{\tilde{\mathrm{T}}:\mathrm{C}} + \mathbf{Z}_{\mathrm{C}}^{\mathrm{out}})^{-1}\mathbf{Z}_{\tilde{\mathrm{T}}:\mathrm{CT}}\,, \tag{A.6}$$

$$\frac{\partial \mathbf{Z}_{\tilde{\mathrm{R}}}^{\mathrm{in}}}{\partial[\boldsymbol{\Psi}]_j} = \mathbf{Z}_{\tilde{\mathrm{R}}:\mathrm{RC}}(\mathbf{Z}_{\tilde{\mathrm{R}}:\mathrm{C}} + \mathbf{Z}_{\mathrm{C}}^{\mathrm{in}})^{-1}\frac{\partial \mathbf{Z}_{\mathrm{C}}^{\mathrm{in}}}{\partial[\boldsymbol{\Psi}]_j}(\mathbf{Z}_{\tilde{\mathrm{R}}:\mathrm{C}} + \mathbf{Z}_{\mathrm{C}}^{\mathrm{in}})^{-1}\mathbf{Z}_{\tilde{\mathrm{R}}:\mathrm{RC}}^{\mathrm{T}}\,, \tag{A.7}$$

$$\begin{aligned}\frac{\partial \mathbf{Z}_{\mathrm{C}}^{\mathrm{out}}}{\partial[\boldsymbol{\Psi}]_j} &= \frac{\partial \mathbf{Z}_{\mathrm{C}:\tilde{\mathrm{T}}}}{\partial[\boldsymbol{\Psi}]_j} - \frac{\partial \mathbf{Z}_{\mathrm{C}:\tilde{\mathrm{R}}\tilde{\mathrm{T}}}^{\mathrm{T}}}{\partial[\boldsymbol{\Psi}]_j}(\mathbf{Z}_{\mathrm{C}:\tilde{\mathrm{R}}} + \mathbf{Z}_{\tilde{\mathrm{R}}}^{\mathrm{out}})^{-1}\mathbf{Z}_{\mathrm{C}:\tilde{\mathrm{R}}\tilde{\mathrm{T}}} \\ &\quad - \mathbf{Z}_{\mathrm{C}:\tilde{\mathrm{R}}\tilde{\mathrm{T}}}^{\mathrm{T}}(\mathbf{Z}_{\mathrm{C}:\tilde{\mathrm{R}}} + \mathbf{Z}_{\tilde{\mathrm{R}}}^{\mathrm{out}})^{-1}\cdot\left[\frac{\partial \mathbf{Z}_{\mathrm{C}:\tilde{\mathrm{R}}\tilde{\mathrm{T}}}}{\partial[\boldsymbol{\Psi}]_j} - \frac{\partial \mathbf{Z}_{\mathrm{C}:\tilde{\mathrm{R}}}}{\partial[\boldsymbol{\Psi}]_j}(\mathbf{Z}_{\mathrm{C}:\tilde{\mathrm{R}}} + \mathbf{Z}_{\tilde{\mathrm{R}}}^{\mathrm{out}})^{-1}\mathbf{Z}_{\mathrm{C}:\tilde{\mathrm{R}}\tilde{\mathrm{T}}}\right],\end{aligned} \tag{A.8}$$

$$\begin{aligned}\frac{\partial \mathbf{Z}_{\mathrm{C}}^{\mathrm{in}}}{\partial[\boldsymbol{\Psi}]_j} &= \frac{\partial \mathbf{Z}_{\mathrm{C}:\tilde{\mathrm{R}}}}{\partial[\boldsymbol{\Psi}]_j} - \frac{\partial \mathbf{Z}_{\mathrm{C}:\tilde{\mathrm{R}}\tilde{\mathrm{T}}}}{\partial[\boldsymbol{\Psi}]_j}(\mathbf{Z}_{\mathrm{C}:\tilde{\mathrm{T}}} + \mathbf{Z}_{\tilde{\mathrm{T}}}^{\mathrm{in}})^{-1}\mathbf{Z}_{\mathrm{C}:\tilde{\mathrm{R}}\tilde{\mathrm{T}}}^{\mathrm{T}} \\ &\quad - \mathbf{Z}_{\mathrm{C}:\tilde{\mathrm{R}}\tilde{\mathrm{T}}}(\mathbf{Z}_{\mathrm{C}:\tilde{\mathrm{T}}} + \mathbf{Z}_{\tilde{\mathrm{T}}}^{\mathrm{in}})^{-1}\cdot\left[\frac{\partial \mathbf{Z}_{\mathrm{C}:\tilde{\mathrm{R}}\tilde{\mathrm{T}}}^{\mathrm{T}}}{\partial[\boldsymbol{\Psi}]_j} - \frac{\partial \mathbf{Z}_{\mathrm{C}:\tilde{\mathrm{T}}}}{\partial[\boldsymbol{\Psi}]_j}(\mathbf{Z}_{\mathrm{C}:\tilde{\mathrm{T}}} + \mathbf{Z}_{\tilde{\mathrm{T}}}^{\mathrm{in}})^{-1}\mathbf{Z}_{\mathrm{C}:\tilde{\mathrm{R}}\tilde{\mathrm{T}}}^{\mathrm{T}}\right].\end{aligned} \tag{A.9}$$

The derivatives of the current noise covariance matrix are the result of changes by the anchor input impedance matrix $\frac{\partial \mathbf{Z}_{\tilde{\mathrm{R}}}^{\mathrm{in}}}{\partial[\boldsymbol{\Psi}]_j}$ and are given by

$$\begin{aligned}\frac{\partial \mathbf{K}}{\partial[\boldsymbol{\Psi}]_j} &= \frac{\partial \mathbf{Y}_{\mathrm{R}}}{\partial[\boldsymbol{\Psi}]_j}(\boldsymbol{\Sigma}^{\mathrm{therm}} + \boldsymbol{\Sigma}^{\mathrm{LNA}})\mathbf{Y}_{\mathrm{R}}^{\mathrm{H}} + \mathbf{Y}_{\mathrm{R}}(\boldsymbol{\Sigma}^{\mathrm{therm}} + \boldsymbol{\Sigma}^{\mathrm{LNA}})\frac{\partial \mathbf{Y}_{\mathrm{R}}^{\mathrm{H}}}{\partial[\boldsymbol{\Psi}]_j} \\ &\quad + \mathbf{Y}_{\mathrm{R}}\left(\frac{\partial \boldsymbol{\Sigma}^{\mathrm{therm}}}{\partial[\boldsymbol{\Psi}]_j} + \frac{\partial \boldsymbol{\Sigma}^{\mathrm{LNA}}}{\partial[\boldsymbol{\Psi}]_j}\right)\mathbf{Y}_{\mathrm{R}}^{\mathrm{H}}\,,\end{aligned} \tag{A.10}$$

$$\frac{\partial \mathbf{Y}_{\mathrm{R}}}{\partial[\boldsymbol{\Psi}]_j} = -\mathbf{Y}_{\mathrm{R}}\frac{\partial \mathbf{Z}_{\tilde{\mathrm{R}}}^{\mathrm{in}}}{\partial[\boldsymbol{\Psi}]_j}\mathbf{Y}_{\mathrm{R}}\,, \tag{A.11}$$

$$\frac{\partial \boldsymbol{\Sigma}^{\mathrm{therm}}}{\partial[\boldsymbol{\Psi}]_j} = 4k_{\mathrm{B}}TB\mathrm{Re}\left(\frac{\partial \mathbf{Z}_{\tilde{\mathrm{R}}}^{\mathrm{in}}}{\partial[\boldsymbol{\Psi}]_j}\right), \tag{A.12}$$

$$\begin{aligned}\frac{\partial \boldsymbol{\Sigma}^{\mathrm{LNA}}}{\partial[\boldsymbol{\Psi}]_j} &= (\sigma^{\mathrm{LNA}})^2\left(\frac{\partial \mathbf{Z}_{\tilde{\mathrm{R}}}^{\mathrm{in}}}{\partial[\boldsymbol{\Psi}]_j}(\mathbf{Z}_{\tilde{\mathrm{R}}}^{\mathrm{in}})^{\mathrm{H}} + \mathbf{Z}_{\tilde{\mathrm{R}}}^{\mathrm{in}}\frac{\partial(\mathbf{Z}_{\tilde{\mathrm{R}}}^{\mathrm{in}})^{\mathrm{H}}}{\partial[\boldsymbol{\Psi}]_j}\right) \\ &\quad - (\sigma^{\mathrm{LNA}})^2 2R^{\mathrm{LNA}}\mathrm{Re}((\rho^{\mathrm{LNA}})^*\frac{\partial \mathbf{Z}_{\tilde{\mathrm{R}}}^{\mathrm{in}}}{\partial[\boldsymbol{\Psi}]_j})\,.\end{aligned} \tag{A.13}$$

The missing non-zero derivatives of $\frac{\partial \mathbf{Z}_{\mathrm{C}:\tilde{\mathrm{T}}}}{\partial[\boldsymbol{\Psi}]_j}$, $\frac{\partial \mathbf{Z}_{\mathrm{C}:\tilde{\mathrm{R}}}}{\partial[\boldsymbol{\Psi}]_j}$ and $\frac{\partial \mathbf{Z}_{\mathrm{C}:\tilde{\mathrm{R}}\tilde{\mathrm{T}}}}{\partial[\boldsymbol{\Psi}]_j}$ follow from

$$\frac{\partial \mathbf{Z}_{\mathrm{C}}}{\partial[\boldsymbol{\Psi}]_j} = j\omega\mathbf{Z}_{\mathrm{C}}(\mathbf{Z}_{\mathrm{C}}^{0})^{-1}\frac{\partial \mathbf{M}}{\partial[\boldsymbol{\Psi}]_j}(\mathbf{Z}_{\mathrm{C}}^{0})^{-1}\mathbf{Z}_{\mathrm{C}} \tag{A.14}$$

and further require the derivatives of all individual mutual inductances $\frac{\partial M_{m,n}}{\partial[\boldsymbol{\Psi}]_j}$ which involve the agents. For the dipole model approximation of the mutual inductance, these have been provided in [26]. In the following, we will also determine them for the

Neumann formula when considering solenoid coils. To this end, we consider $[\mathbf{\Psi}]_j$ to be one of the deployment parameters of coil m. Consequently, we apply the Leibniz integral rule to (2.2), which results in

$$\frac{\partial M_{m,n}}{\partial[\mathbf{\Psi}]_j} = K^{\text{Neu}} \oint_{\text{C}_\text{n}} \oint_{\text{C}_\text{m}} \frac{\partial}{\partial[\mathbf{\Psi}]_j} \left(\frac{\text{d}\mathbf{l}_m}{d^{\text{wire}}(l_m, l_n)} \right) \text{d}\mathbf{l}_n, \tag{A.15}$$

$$= K^{\text{Neu}} \oint_{\text{C}_\text{n}} \oint_{\text{C}_\text{m}} \frac{\frac{\partial \text{d}\mathbf{l}_m}{\partial[\mathbf{\Psi}]_j} \text{d}\mathbf{l}_n}{d^{\text{wire}}(l_m, l_n)} - \frac{\text{d}\mathbf{l}_m \text{d}\mathbf{l}_n \frac{\partial d^{\text{wire}}(l_m, l_n)}{\partial[\mathbf{\Psi}]_j}}{(d^{\text{wire}})^2(l_m, l_n)} \tag{A.16}$$

The vectorial line-segment $\text{d}\mathbf{l}_m$ is defined by the given wire positions $\mathbf{p}_m^{\text{wire}}(l_m)$ at length $l_m \in [0, l^{\text{wire}}[$, where l^{wire} is the total wire length. Using a coordinate frame transformation, this wire position can generally be expressed as

$$\mathbf{p}_m^{\text{wire}}(l_m) = \mathbf{p}_m + \mathbf{O}_m \, \mathbf{p}_{\text{orig}}(l_m) \tag{A.17}$$

where $\mathbf{O}_m$ is the three-dimensional rotation matrix according to the respective coil's orientation given by

$$\mathbf{O}_m = \mathbf{O}_z(\alpha_m) \, \mathbf{O}_y(\beta_m) \, \mathbf{O}_z(\gamma_m) \tag{A.18}$$

$$= \begin{bmatrix} \cos(\alpha_m) & -\sin(\alpha_m) & 0 \\ \sin(\alpha_m) & \cos(\alpha_m) & 0 \\ 0 & 0 & 1 \end{bmatrix} \begin{bmatrix} \cos(\beta_m) & 0 & \sin(\beta_m) \\ 0 & 1 & 0 \\ -\sin(\beta_m) & 0 & \cos(\beta_m) \end{bmatrix} \begin{bmatrix} \cos(\gamma_m) & -\sin(\gamma_m) & 0 \\ \sin(\gamma_m) & \cos(\gamma_m) & 0 \\ 0 & 0 & 1 \end{bmatrix}. \tag{A.19}$$

The deployment independent function $\mathbf{p}_{\text{orig}}(l_m)$ is the position formula of an upright solenoid in the coordinate origin as shown in Fig. 2.1 and its spiral part can be expressed as

$$\mathbf{p}_{\text{orig}}(l_m) = \begin{bmatrix} D^{\text{coil}} \cos\left(2\pi N^{\text{coil}} \frac{l_m}{l^{\text{spiral}}}\right) \\ D^{\text{coil}} \sin\left(2\pi N^{\text{coil}} \frac{l_m}{l^{\text{spiral}}}\right) \\ H^{\text{coil}} \frac{l_m}{l^{\text{spiral}}} - \frac{H^{\text{coil}}}{2} \end{bmatrix}, \quad 0 \le l_m \le l^{\text{spiral}} \tag{A.20}$$

with coil radius D^{coil}, number of turns N^{coil}, total height H^{coil} and $l^{\text{spiral}} < l^{\text{wire}}$ as the total length of the coil's spiral part. The remaining part of the coil is a straight connector on both ends with an offset of $\frac{H^{\text{coil}} - D^{\text{wire}}}{N^{\text{coil}}}$ from both spiral ends. With this geometric restriction, the vectorial line-segment of the spiral part can be formulated as

$$\mathrm{d}\mathbf{l}_m = \frac{\partial \mathbf{p}_m^{\mathrm{wire}}(l_m)}{\partial l_m} \| \frac{\partial \mathbf{p}_m^{\mathrm{wire}}(l_m)}{\partial l_m} \|^{-1} \mathrm{d}l_m \tag{A.21}$$

$$= \mathbf{O}_m \frac{2\pi N^{\mathrm{coil}} \mathrm{d}l_m}{\sqrt{(2\pi N^{\mathrm{coil}} D^{\mathrm{coil}})^2 + (H^{\mathrm{coil}})^2}} \begin{bmatrix} -D^{\mathrm{coil}} \sin\left(2\pi N^{\mathrm{coil}} \frac{l_m}{l^{\mathrm{spiral}}}\right) \\ D^{\mathrm{coil}} \cos\left(2\pi N^{\mathrm{coil}} \frac{l_m}{l^{\mathrm{spiral}}}\right) \\ \frac{H^{\mathrm{coil}}}{2\pi N^{\mathrm{coil}}} \end{bmatrix}. \tag{A.22}$$

For the coil's connector, it is simply a unit vector of the corresponding direction. The required derivative of the vectorial line-segment is only non-zero for derivatives of the orientational parameters and follows as

$$\frac{\partial(\mathrm{d}\mathbf{l}_m)}{\partial[\mathbf{\Psi}]_j} = \frac{\partial \mathbf{O}_m}{\partial[\mathbf{\Psi}]_j} \mathbf{O}_m^{-1} \mathrm{d}\mathbf{l}_m \,. \tag{A.23}$$

Lastly, for the derivative of the distance between any given pair of wire elements we have

$$\frac{\partial d^{\mathrm{wire}}(l_m, l_n)}{\partial[\mathbf{\Psi}]_j} = \frac{(\mathbf{p}_m^{\mathrm{wire}}(l_m) - \mathbf{p}_n^{\mathrm{wire}}(l_n))}{d^{\mathrm{wire}}(l_m, l_n)} \frac{\partial \mathbf{p}_m^{\mathrm{wire}}(l_m)}{\partial[\mathbf{\Psi}]_j}. \tag{A.24}$$

with $\frac{\partial \mathbf{p}_m^{\mathrm{wire}}(l_m)}{\partial[\mathbf{\Psi}]_j}$ being the corresponding unit direction vector in case of spatial derivatives or $\frac{\partial \mathbf{O}_m}{\partial[\mathbf{\Psi}]_j}\mathbf{p}_{\mathrm{orig}}(l_m)$ in case of orientational derivatives. The derivatives of the rotation matrix are obtained via basic trigonometric derivatives of (A.18). We can hence numerically evaluate the derivative of the mutual inductance (A.16) according to the Neumann formula for any arbitrary solenoid coil pair (m, n).

A.1 Relay-Specific Derivatives

As mentioned in Sec. 5.1, the impact of passive relays can easily be included into the existing model by replacing the previous impedance matrix $\mathbf{Z}_\mathrm{C}$ of the agent-anchor coil antennas by the matrix $\tilde{\mathbf{Z}}_\mathrm{C}$, which is the relay incorporated impedance matrix of the agent-anchor coupling. This extension requires the derivatives of (5.2), which are found as

$$\begin{aligned}\frac{\partial \tilde{\mathbf{Z}}_{\mathrm{C}}}{\partial[\boldsymbol{\Psi}]_j} = \frac{\partial \mathbf{Z}_{\mathrm{C}}}{\partial[\boldsymbol{\Psi}]_j} &- \frac{\partial\begin{bmatrix}\mathbf{Z}_{\mathrm{C:YT}}^{\mathrm{T}}\\ \mathbf{Z}_{\mathrm{C:YR}}^{\mathrm{T}}\end{bmatrix}}{\partial[\boldsymbol{\Psi}]_j}\left(\mathbf{Z}_{\mathrm{C:Y}} + \mathbf{Z}_{\mathrm{L}}\right)^{-1}\begin{bmatrix}\mathbf{Z}_{\mathrm{C:YT}} & \mathbf{Z}_{\mathrm{C:YR}}\end{bmatrix}\\ &+ \begin{bmatrix}\mathbf{Z}_{\mathrm{C:YT}}^{\mathrm{T}}\\ \mathbf{Z}_{\mathrm{C:YR}}^{\mathrm{T}}\end{bmatrix}\left(\mathbf{Z}_{\mathrm{C:Y}} + \mathbf{Z}_{\mathrm{L}}\right)^{-1}\frac{\partial \mathbf{Z}_{\mathrm{C:Y}}}{\partial[\boldsymbol{\Psi}]_j}\left(\mathbf{Z}_{\mathrm{C:Y}} + \mathbf{Z}_{\mathrm{L}}\right)^{-1}\begin{bmatrix}\mathbf{Z}_{\mathrm{C:YT}} & \mathbf{Z}_{\mathrm{C:YR}}\end{bmatrix}\\ &- \begin{bmatrix}\mathbf{Z}_{\mathrm{C:YT}}^{\mathrm{T}}\\ \mathbf{Z}_{\mathrm{C:YR}}^{\mathrm{T}}\end{bmatrix}\left(\mathbf{Z}_{\mathrm{C:Y}} + \mathbf{Z}_{\mathrm{L}}\right)^{-1}\frac{\partial\begin{bmatrix}\mathbf{Z}_{\mathrm{C:YT}} & \mathbf{Z}_{\mathrm{C:YR}}\end{bmatrix}}{\partial[\boldsymbol{\Psi}]_j}.\end{aligned} \tag{A.25}$$

The calculation of this quantity again requires the derivatives of all agent-involving mutual inductances $\frac{\partial M_{m,n}}{\partial[\boldsymbol{\Psi}]_j}$.

Appendix B

Trace Maximization for Pairwise Distance Estimates

The ML distance and orientation estimator for a pair of three-axis coils requires the maximization of the trace in (6.25). We replace $\mathbf{B}_{m,n}$ with its constituents and drop the indices m, n, i.e. $\mathrm{tr}\left(\mathbf{A}_{m,n}\mathbf{B}_{m,n}\right) = \mathrm{tr}\left(\mathbf{AFO}\right)$. The goal is to maximize this trace with the underlying constraints

$$\mathbf{O}\mathbf{O}^{\mathrm{T}} = 1\,, \qquad \det(\mathbf{O}) = 1\,, \qquad \mathbf{F} = (\frac{3}{2}\mathbf{u}\mathbf{u}^{\mathrm{T}} - \frac{1}{2}\mathbf{I}_3)\,, \qquad \|\mathbf{u}\| = 1\,.$$

Applying **S**ingular **V**alue **D**ecompositions (SVDs) as well as an **E**igen(**V**alue)**D**ecomposition (EVD) we can rewrite the trace as

$$\mathrm{tr}\left(\mathbf{AFO}\right) = \mathrm{tr}\left(\overbrace{\mathbf{U}_A\mathbf{S}_A\mathbf{V}_A^{\mathrm{T}} \cdot \mathbf{U}_F \underbrace{\mathbf{\Sigma}_F}_{\mathbf{S}_F\mathbf{C}_F} \mathbf{U}_F^{\mathrm{T}}}^{\mathbf{U}_{AF}\mathbf{S}_{AF}\mathbf{V}_{AF}^{\mathrm{T}}} \cdot \mathbf{U}_O\mathbf{V}_O^{\mathrm{T}}\right), \tag{B.1}$$

with $\mathbf{S}_F = \mathrm{diag}(1,\, \frac{1}{2},\, \frac{1}{2})$ and $\mathbf{C}_F = \mathrm{diag}(1,\, -1,\, -1)$ due to the structural constraints on $\mathbf{F}$, which hence leads to $\det(\mathbf{U}_F\mathbf{C}_F\mathbf{U}_F^{\mathrm{T}}) = 1$ for any possible choice. Note that all corresponding singular and eigenvalues are always considered in descending order.

From [150] we further know that for any choice of $\mathbf{F}$, the optimal agent orientation matrix is given by

$$\mathbf{O}_{\mathrm{opt}} = \mathbf{V}_{AF} \underbrace{\mathrm{diag}(1,\, 1,\, \det(\mathbf{U}_A\mathbf{V}_A^{\mathrm{T}}))}_{\mathbf{E}_A} \mathbf{U}_{AF}^{\mathrm{T}} \quad \text{if} \quad \det(\mathbf{O}) \overset{!}{=} 1\,, \tag{B.2}$$

since $\det(\mathbf{U}_A\mathbf{V}_A^{\mathrm{T}}) = \det(\mathbf{U}_{AF}\mathbf{V}_{AF}^{\mathrm{T}})$. This also implies that the upper bound according to the Von Neumann trace inequality cannot be obtained due to our constraints on $\mathbf{O}$ if $\det(\mathbf{U}_A\mathbf{V}_A^{\mathrm{T}}) = -1$. Furthermore, the best feasible realization of $\mathbf{F}$ leads to $\mathbf{S}_{AF} = \mathbf{S}_A\mathbf{S}_F$ and we hence find

$$\mathrm{tr}\left(\mathbf{AFO}\right) \overset{(a)}{\leq} \mathrm{tr}\left(\mathbf{S}_{AF}\mathbf{E}_A\right) \overset{(b)}{\leq} \mathrm{tr}\left(\mathbf{S}_A\mathbf{S}_F\mathbf{E}_A\right) \overset{(c)}{\leq} \mathrm{tr}\left(\mathbf{S}_A\mathbf{S}_F\right). \tag{B.3}$$

With an optimal constrained solution

$$\hat{\mathbf{O}} = \mathbf{V}_A \mathbf{C}_F \mathbf{E}_A \mathbf{U}_A^{\mathrm{T}}, \tag{B.4}$$

$$\hat{\mathbf{F}} = \mathbf{V}_A \mathbf{S}_F \mathbf{C}_F \mathbf{V}_A^{\mathrm{T}}, \tag{B.5}$$

we attain equality in (a) and (b) via (B.4) and (B.5). Moreover, under our constraints these choices also yield equality in (c) if and only if $\det(\mathbf{U}_A \mathbf{V}_A^{\mathrm{T}}) = 1$ and for this case coincide with the optimal unconstrained solution.

Appendix C

Preliminary Measurements of an Anchor-Tag Coil Pair

This appendix summarizes preliminary measurement findings regarding the detuning of solenoid coils and the associated noise statistics of the Rohde & Schwarz ZNBT8 VNA, which was used throughout this work. The VNA is used to measure the reflection parameter of the anchor coils and automatically transforms them to the corresponding complex input impedance. In this preliminary study, we only use the coil antennas of Tab. 9.1 without any additional electronics such as matching networks or switches. The goal of this appendix is to obtain general information on extent of the detuning, to analyze the observed noise statistics, and to assess the influence of environmental clutter.

In detail, we use the hip coil (cf. Tab. 9.1) as an anchor and measure its input impedance $Z^{\mathrm{in}}_{\mathrm{prelim}}$ via $N_{\mathrm{prelim}} = 256$ frequency sweeps in the range $f \in [400, 600]$ kHz. Simultaneously, the right thigh coil acts as a resonantly loaded passive tag and is placed coaxially in the anchor's vicinity at variable coil pair distances d^{pair}. In some realizations, we also place a possibly distorting object on the opposite side of the anchor at a distance d^{obj}, as illustrated in Fig. C.1. Generally, the input impedance $Z^{\mathrm{in}}_{\mathrm{prelim}}(d^{\mathrm{pair}}, d^{\mathrm{obj}}, f)$ is hence a function of both distances and the frequency. For notational convenience, we drop the corresponding parameters for the special cases $d^{\mathrm{obj}} \to \infty$ (no distorting object), $d^{\mathrm{pair}} = 35\,\mathrm{cm}$ (reference distance of the tag), or $f = 500\,\mathrm{kHz}$ (resonance frequency of the individual tag). The quantity $Z^{\mathrm{in}}_{\mathrm{prelim}}$ without any specifying parameters is thus the measured input impedance at $f = 500\,\mathrm{kHz}$ if only the passive tag is present and located at a distance of $d^{\mathrm{pair}} = 35\,\mathrm{cm}$.

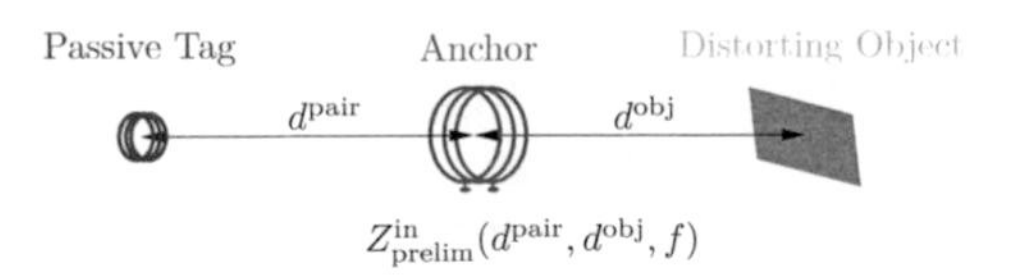

Figure C.1: Illustration of the preliminary measurement setup containing a single coaxially aligned anchor-tag pair with inter-coil distance d^{pair} and possibly different distorting objects at distance d^{obj} to the anchor.

C.1 General Detuning Observations for an Anchor-Tag Pair

In a first instance, we look at a scenario where only the anchor and the passive tag are present, i.e. $d^{\mathrm{obj}} \to \infty$. Fig. C.2a summarizes the measurements for different inter-coil distances and frequencies as locus plot. It reveals that the reactance is at least an order of magnitude higher than the resistance and increases for higher frequencies. Moreover, the presence of the single passive relay leads to a resonance peak close to 500 kHz for the resistance and a point of inflection (with a zero crossing) for the reactance (cf. Fig. C.2c). This behavior is understood by looking at the mathematical description of the input impedance (ignoring the self capacitances):

$$Z^{\mathrm{in}}_{\mathrm{prelim}}(d^{\mathrm{pair}}, f) = R_{n_{\mathrm{R}}} + j\omega L_{n_{\mathrm{R}}} + \omega^2 M^2_{n_{\mathrm{R}},n_{\mathrm{T}}} \left(R_{n_{\mathrm{T}}} + j\omega L_{n_{\mathrm{T}}} - j\frac{(\omega^{\mathrm{des}})^2 L_{n_{\mathrm{T}}}}{\omega} \right)^{-1} \tag{C.1}$$

$$= R_{n_{\mathrm{R}}} + j\omega L_{n_{\mathrm{R}}} + \omega^3 M^2_{n_{\mathrm{R}},n_{\mathrm{T}}} \frac{R_{n_{\mathrm{T}}}\omega + jL_{n_{\mathrm{T}}}((\omega^{\mathrm{des}})^2 - \omega^2)}{R^2_{n_{\mathrm{T}}}\omega^2 + L^2_{n_{\mathrm{T}}}((\omega^{\mathrm{des}})^2 - \omega^2)^2}, \tag{C.2}$$

where $\omega^{\mathrm{des}} = 2\pi f^{\mathrm{des}} = 2\pi \cdot 500\,\mathrm{kHz}$ is the design frequency of the passive tag. The entire impact of the passive tag is the impedance difference $\Delta Z^{\mathrm{in}}_{\mathrm{prelim}}(d^{\mathrm{pair}}, f) = Z^{\mathrm{in}}_{\mathrm{prelim}}(d^{\mathrm{pair}}, f) - Z^{\mathrm{in}}_{\mathrm{prelim}}(d^{\mathrm{pair}} \to \infty, f)$ and can be approximated by the last summand of (C.2). Clearly, we can determine the location of the resonance peak, i.e. a maximum of this function, by finding the angular frequency for which the derivative of its real part $\mathrm{Re}\left(\Delta Z^{\mathrm{in}}_{\mathrm{prelim}}(d^{\mathrm{pair}}, f)\right)$ is zero. Via simple derivation rules, this location follows as $\omega_0 = \sqrt{\frac{2(\omega^{\mathrm{des}})^4 L^2_{n_{\mathrm{T}}}}{2(\omega^{\mathrm{des}})^2 L^2_{n_{\mathrm{T}}} - R^2_{n_{\mathrm{T}}}}} = \omega^{\mathrm{des}} \left(1 - \frac{R^2_{n_{\mathrm{T}}}}{2(\omega^{\mathrm{des}})^2 L^2_{n_{\mathrm{T}}}}\right)^{-\frac{1}{2}} \overset{R_{n_{\mathrm{T}}} \to 0}{=} \omega^{\mathrm{des}}$. We can further repeat this step for the imaginary part $\mathrm{Im}\left(\Delta Z^{\mathrm{in}}_{\mathrm{prelim}}(d^{\mathrm{pair}}, f)\right)$ and find the theoretical location of the the point of inflection at $\omega^{\mathrm{infl}} = \omega^{\mathrm{des}}$. The corresponding measurements of this impedance difference are shown in Fig. C.2b and Fig. C.2c. The magnitude of this impedance difference clearly decreases rapidly over distance and approaches a noise floor at roughly $1 \times 10^{-2}\,\Omega$, which is apparent from Fig. C.2c. Consequently, even if the tag impact is measured at the correct resonance frequency, it is not reliably detectable anymore for distance larger than $d^{\mathrm{pair}} \approx 70\,\mathrm{cm}$.

These results highlight the range limitation of the passive MI links and reinforce our prior assumptions that it is helpful to use multiple anchors which are placed as close as possible to the different limbs. This causes an anchor to be mostly respondent to the passive tags in its direct vicinity, e.g. a shoulder coil is mostly affected by

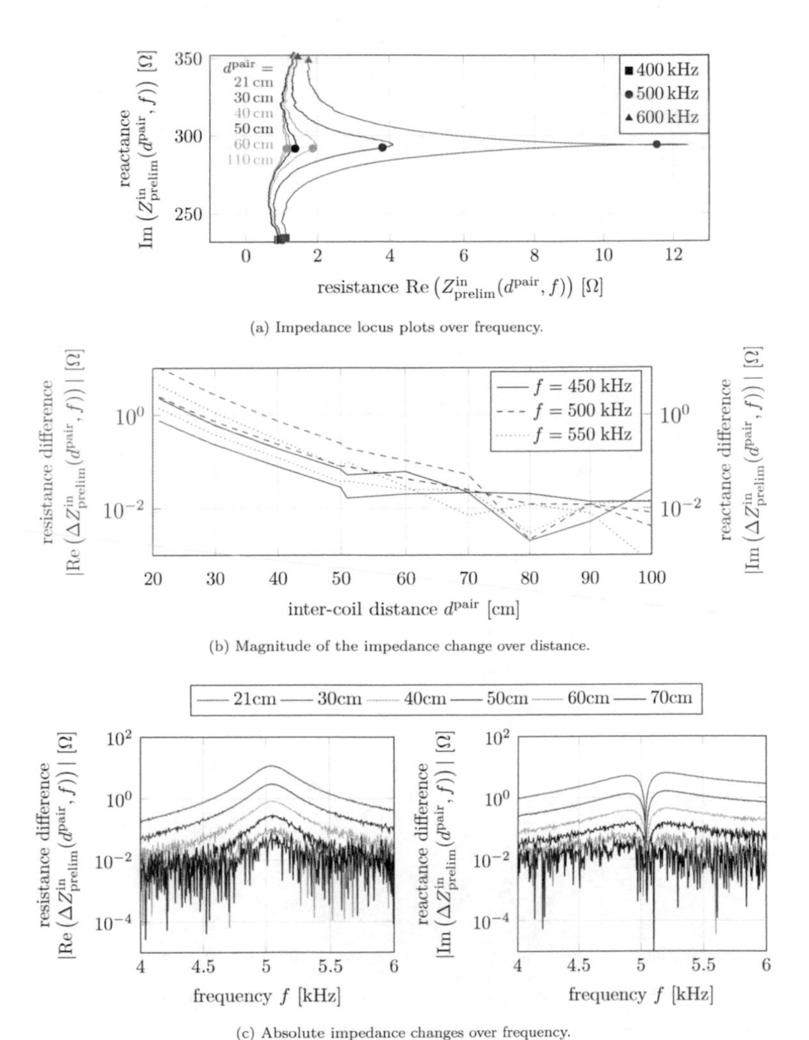

(a) Impedance locus plots over frequency.

(b) Magnitude of the impedance change over distance.

(c) Absolute impedance changes over frequency.

Figure C.2: Various illustrations of the frequency and distance behavior for the input impedance $Z^{\mathrm{in}}_{\mathrm{prelim}}(d^{\mathrm{pair}}, f)$ and the impedance difference $\Delta Z^{\mathrm{in}}_{\mathrm{prelim}}(d^{\mathrm{pair}}, f)$ of the coaxially aligned anchor-tag coil pair without nearby distorting objects.

the passive tags of the corresponding arm and head. Vice versa, it also indicates that the system may be generally unaffected by neighboring systems, if they are not in the direct vicinity. However, due to the presence and mutual coupling between multiple passive tags, the overall detuning of the full demonstrator may sometimes be noticeable even for ranges larger than 70 cm, e.g. if the combination of passive tags causes a waveguide effect. Moreover, with multiple passive tags the resonance peak might split (resonance mode splitting) and the resulting resonance frequencies might be shifted more drastically. Lastly, the observed noise floor is unlikely to be fixed but rather dependent of the transmit power, other device-specific parameters and the measurement method in general.

C.2 Impact of Distorting Objects

Next, we look at the impact that different objects have on the impedance measurements. In the simulation-based study we already examined the impact of a second neighboring resonant system and found such a system to be negligible for larger distances. Moreover, the impact of a neighboring system may be further reduced if close-by systems communicate and share the medium in a sophisticated manner to avoid interference and distortions. For environmental clutter, this is however unfeasible and the resulting irregular distortions are hard to mitigate or account for. As a result, they may severely reduce the classification capabilities of the planned MI system. Apart from active interference of external magnetic fields, MI systems are usually distorted if conducting materials are nearby. To this end, we define a relative impedance difference η^{clutter}, with its real part

$$\mathrm{Re}\left(\eta^{\text{clutter}}(d^{\text{obj}})\right) = \frac{\mathrm{Re}\left(Z^{\text{in}}_{\text{prelim}}(d^{\text{obj}}) - Z^{\text{in}}_{\text{prelim}}\right)}{\mathrm{Re}\left(Z^{\text{in}}_{\text{prelim}} - Z^{\text{in}}_{\text{prelim}}(d^{\text{pair}} \to \infty)\right)}, \tag{C.3}$$

and an analogously defined imaginary part. This quantity, which is shown in Fig. C.3a, represents the additional impedance change caused by the nearby object, relative to the change that is already caused by the passive tag at $d^{\text{pair}} = 35$ cm. It can be observed that its real part is almost unaffected by clutter, whereas the imaginary part suffers from distortions almost regardless of which object is present. The strongest impact can be seen for a ferromagnetic steel plate and a monitor, which both cause significant distortions up to a distance of $d^{\text{obj}} = 60$ cm. However, even the other objects cause distortions on the imaginary part that amount to 10% at a distance of

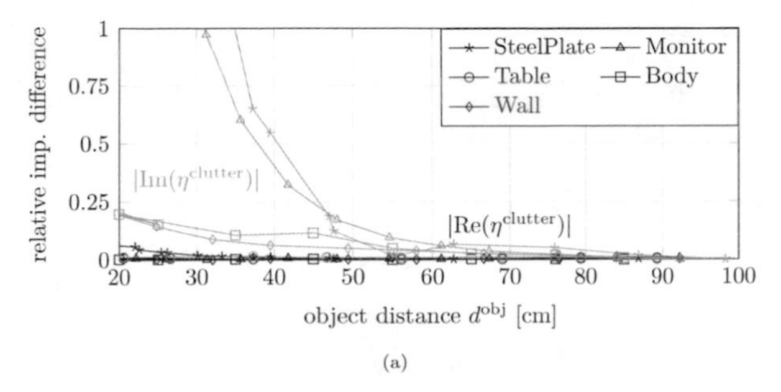

(a)

Figure C.3: Magnitudes of the real and imaginary parts of the relative impedance difference $\eta^{clutter}(d^{obj})$ for a coaxially-aligned anchor-tag pair with different distorting objects at distance d^{obj}.

$d^{obj} = 35\,\mathrm{cm}$. Overall, we infer that the posture recognition system is indeed impacted by environmental clutter and its performance may degrade when being too close to large ferromagnetic objects. If the system is to be used in a distorting environment, additional distortion mitigation or rejection approaches (cf. [25]) may be beneficial. However, those approaches are out of our scope and the performance analyses of this work assume an environment without environmental clutter. Moreover, minor distortions caused by the human body itself are accounted for by the training process of the supervised classifiers.

C.3 Noise Characterization

In a next step, we want to characterize the noise statistics of the setup by further analyzing the measurement samples of the coaxially aligned anchor-tag pair without distorting objects. Fig. C.4a hence shows the mean of the input impedance magnitude $\mu(|Z^{in}_{prelim}(d^{pair}, f)|)$ over all samples n for each different frequency. The related dispersion index, which is the ratio of the observed variance and mean $\frac{\sigma^2(|Z^{in}_{prelim}(d^{pair},f)|)}{\mu(|Z^{in}_{prelim}(d^{pair},f)|)}$ is also displayed. Instead of being constant, which would implicate a proportional relationship between measurement magnitude and noise variance, the dispersion index increases approximately linearly. This observation implicates colored instead of white noise, which may be related to frequency dependencies of the measurement device's electronics. However, even at higher frequencies, the measured dispersion index is still significantly lower than 8×10^{-4}, which was the lowest dispersion index used for the AWGN of the simulation-based study (cf. Sec. 8.5).

Next, we look at the residual measurement errors $\epsilon_n(d^{pair}, f) = Z^{in}_{prelim,n}(d^{pair}, f) -$

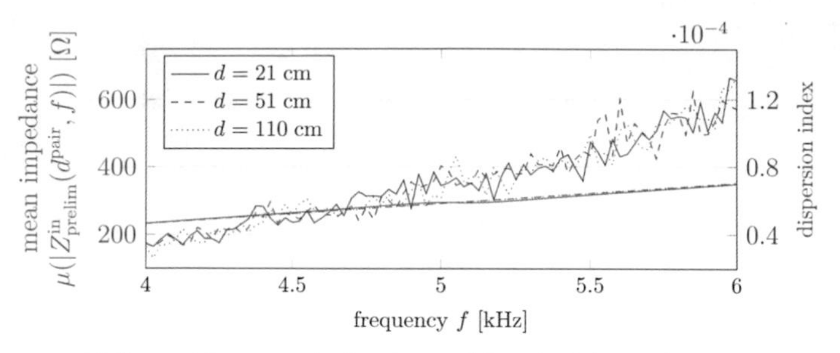

(a) Change of the mean anchor impedance and dispersion index over frequency.

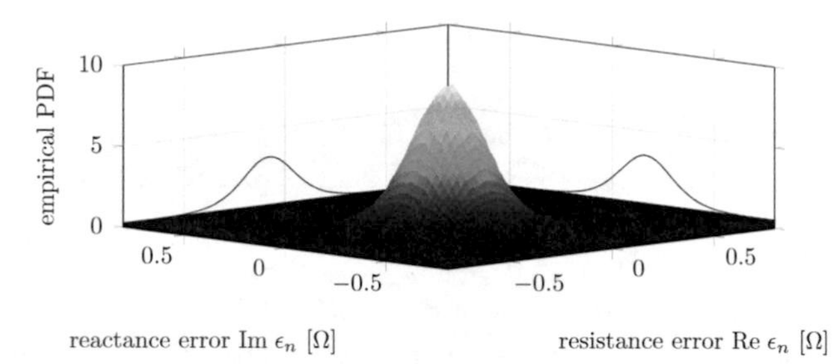

(b) Empirical joint and marginal PDFs of the errors ϵ_n for all distances and frequencies.

Figure C.4: Empirical measurement error characterization for a coaxially aligned anchor-tag pair for various distances and frequencies.

$\mu(Z^{\text{in}}_{\text{prelim}}(d^{\text{pair}}, f))$. In Fig. C.4b we show the joint and marginal PDFs of these measurement errors over all frequencies and distances. The shape of the joint PDF resembles a two-dimensional normal distribution with the same variance for both the resistance errors and reactance errors. The same behavior was also observed when only considering the single-distance or single-frequency measurements (not shown).

Mathematical Notation and Operators

Notation	Name	Description
a, A	scalar	cursive lower or upper case letter
$\mathbf{a}$	column vector	lower case boldface letter
$\mathbf{A}$	matrix	upper case boldface letter
$[\mathbf{A}]_{m,:}$	matrix row	row m of matrix $\mathbf{A}$
$[\mathbf{A}]_{:,n}$	matrix column	column n of matrix $\mathbf{A}$
$[\mathbf{A}]_{m,n}$	matrix element	matrix element of the m-th row and n-th column
$(.)^{\mathrm{T}}$	transpose	transpose of vector or matrix
$(.)^{*}$	complex conj.	complex conjugate of scalar, vector, or matrix
$(.)^{\mathrm{H}}$	conj. transpose	conjugate transpose of vector or matrix
$(.)^{-1}$	inverse	inverse of matrix
$\mathbf{U}_A \mathbf{S}_A \mathbf{V}_A^{\mathrm{H}}$	SVD	singular value decomposition of matrix $\mathbf{A}$
$\mathbf{U}_A \mathbf{\Lambda}_A \mathbf{U}_A^{-1}$	EVD	eigendecomposition of full rank square matrix $\mathbf{A}$
$\mathbf{I}_N$	unit matrix	unit matrix of dimension $N \times N$
$\mathbf{1}_{M \times N}$	all-ones matrix	all-ones matrix of dimension $M \times N$
$\mathbf{0}_{M \times N}$	all-zero matrix	all-zero matrix of dimension $M \times N$
j	imaginary unit	imaginary unit which fulfills $j^2 = -1$
$\mathrm{Re}(\cdot)$	real part	real part
$\mathrm{Im}(\cdot)$	imaginary part	imaginary part
$\mathrm{sgn}(\cdot)$	sign	sign function
$\mathrm{tr}(\cdot)$	trace	trace of matrix
$\det(\cdot)$	determinant	determinant of matrix
$\mathrm{vec}(\cdot)$	vectorization	vectorization of matrix
$\mathrm{diag}(\cdot)$	diagonalization	diagonal matrix
$\hat{(\cdot)}$	estimate	estimate of scalar, vector, or matrix
$\|(\cdot)\|$	Euclidean norm	Euclidean norm
$\|(\cdot)\|_F$	Frobenius norm	Frobenius norm
$(\cdot) \parallel (\cdot)$	parallel	parallel connection of two impedances
$f_a(a)$	PDF	probability density function of random variable a
$F_a(a)$	CDF	cumulative distribution function of random variable a

List of Acronyms

AWGN	Additive White Gaussian Noise
CDF	Cumulative Distribution Function
CNN	Convolutional Neural Network
CPU	Central Processing Unit
CRLB	Cramér-Rao Lower Bound
DPEB	Directional Position Error Bound
EVD	Eigen(Value)Decomposition
FIM	Fisher Information Matrix
GNSS	Global Navigation Satellite System
IMU	Inertial Measurement Unit
IoT	Internet of Things
kNN	k-Nearest Neighbors
LHF	LikeliHood Function
LNA	Low-Noise Amplifier
MAP	Maximum A Posteriori
MI	Magneto-Inductive
MIMO	Multiple Input Multiple Output
ML	Maximum Likelihood
MLP	MultiLayer Perceptron
MQS	MagnetoQuasiStatic
MSD	MusculoSkeletal Disorder
PCA	Principal Component Analysis
PCB	Printed Circuit Board
PDF	Probability Density Function
PEB	Position Error Bound
RMSE	Root-Mean-Square Error
SIMO	Single Input Multiple Output
SLS	Single-Layer Solenoid
SMA	SubMiniature version A
SNR	Signal-to-Noise Ratio
SVD	Singular Value Decomposition

SVM	Support Vector Machine
tSNE	t-distributed Stochastic Neighbor Embedding
UWB	Ultra-WideBand
VNA	Vector Network Analyzer
WLAN	Wireless Local Area Network
WPT	Wireless Power Transfer

Bibliography

[1] P. Cerwall and et al., "Ericsson mobility report," *June*, 2021.

[2] S. Al-Sarawi, M. Anbar, R. Abdullah, and A. B. Al Hawari, "Internet of things market analysis forecasts, 2020–2030," in *2020 Fourth World Conference on smart trends in systems, security and sustainability (WorldS4)*. IEEE, 2020, pp. 449–453.

[3] J. Cancela, I. Charlafti, S. Colloud, and C. Wu, "Digital health in the era of personalized healthcare: opportunities and challenges for bringing research and patient care to a new level," *Digital Health*, pp. 7–31, 2021.

[4] A. Rizwan, A. Zoha, R. Zhang, W. Ahmad, K. Arshad, N. A. Ali, A. Alomainy, M. A. Imran, and Q. H. Abbasi, "A review on the role of nano-communication in future healthcare systems: A big data analytics perspective," *IEEE Access*, vol. 6, pp. 41 903–41 920, 2018.

[5] H. Ceylan, I. C. Yasa, U. Kilic, W. Hu, and M. Sitti, "Translational prospects of untethered medical microrobots," *Progress in Biomedical Engineering*, vol. 1, no. 1, p. 012002, 2019.

[6] C. Steiger, A. Abramson, P. Nadeau, A. P. Chandrakasan, R. Langer, and G. Traverso, "Ingestible electronics for diagnostics and therapy," *Nature Reviews Materials*, vol. 4, no. 2, pp. 83–98, 2019.

[7] J. Grosinger, W. Pachler, and W. Bosch, "Tag size matters: Miniaturized RFID tags to connect smart objects to the internet," *IEEE microwave magazine*, vol. 19, no. 6, pp. 101–111, 2018.

[8] M. Vallejo, J. Recas, P. G. Del Valle, and J. L. Ayala, "Accurate human tissue characterization for energy-efficient wireless on-body communications," *Sensors*, vol. 13, no. 6, pp. 7546–7569, 2013.

[9] M. Salayma, A. Al-Dubai, I. Romdhani, and Y. Nasser, "Wireless body area network (WBAN) a survey on reliability, fault tolerance, and technologies coexistence," *ACM Computing Surveys (CSUR)*, vol. 50, no. 1, pp. 1–38, 2017.

[10] R. Du, A. Ozcelikkale, C. Fischione, and M. Xiao, "Optimal energy beamforming and data routing for immortal wireless sensor networks," in *2017 IEEE International Conference on Communications (ICC)*. IEEE, 2017, pp. 1–6.

[11] S. Niu, N. Matsuhisa, L. Beker, J. Li, S. Wang, J. Wang, Y. Jiang, X. Yan, Y. Yun, W. Burnett *et al.*, "A wireless body area sensor network based on stretchable passive tags," *Nature Electronics*, vol. 2, no. 8, pp. 361–368, 2019.

[12] S. Bi, C. K. Ho, and R. Zhang, "Wireless powered communication: Opportunities and challenges," *IEEE Communications Magazine*, vol. 53, no. 4, pp. 117–125, 2015.

[13] Y. Wang, H. Wang, J. Xuan, and D. Y. Leung, "Powering future body sensor network systems: A review of power sources," *Biosensors and Bioelectronics*, vol. 166, p. 112410, 2020.

[14] S. Movassaghi, M. Abolhasan, J. Lipman, D. Smith, and A. Jamalipour, "Wireless body area networks: A survey," *IEEE Communications surveys & tutorials*, vol. 16, no. 3, pp. 1658–1686, 2014.

[15] S. Ullah, H. Higgins, B. Braem, B. Latre, C. Blondia, I. Moerman, S. Saleem, Z. Rahman, and K. S. Kwak, "A comprehensive survey of wireless body area networks," *Journal of medical systems*, vol. 36, no. 3, pp. 1065–1094, 2012.

[16] S. Basu, M. Sarkar, S. Nagaraj, and S. Chinara, "A survey on ultra wideband and ultrasonic communication for body area networks," *International Journal of Ultra Wideband Communications and Systems*, vol. 3, no. 3, pp. 143–154, 2016.

[17] L. Galluccio, T. Melodia, S. Palazzo, and G. E. Santagati, "Challenges and implications of using ultrasonic communications in intra-body area networks," in *2012 9th Annual Conference on Wireless On-Demand Network Systems and Services (WONS)*. IEEE, 2012, pp. 182–189.

[18] M. Li and Y. T. Kim, "Feasibility analysis on the use of ultrasonic communications for body sensor networks," *Sensors*, vol. 18, no. 12, p. 4496, 2018.

[19] G. Dumphart, B. I. Bitachon, and A. Wittneben, "Magneto-inductive powering and uplink of in-body microsensors: Feasibility and high-density effects," in *2019 IEEE WCNC*. IEEE, 2019, pp. 1–6.

[20] E. Reusens, W. Joseph, B. Latré, B. Braem, G. Vermeeren, E. Tanghe, L. Martens, I. Moerman, and C. Blondia, "Characterization of on-body communication channel and energy efficient topology design for wireless body area networks," *IEEE Transactions on Information Technology in Biomedicine*, vol. 13, no. 6, pp. 933–945, 2009.

[21] R. M. Shubair and H. Elayan, "In vivo wireless body communications: State-of-the-art and future directions," in *2015 Loughborough Antennas & Propagation Conference (LAPC)*. IEEE, 2015, pp. 1–5.

[22] M. Murad, A. A. Sheikh, M. A. Manzoor, E. Felemban, and S. Qaisar, "A survey on current underwater acoustic sensor network applications," *International Journal of Computer Theory and Engineering*, vol. 7, no. 1, p. 51, 2015.

[23] Y. Li, S. Wang, C. Jin, Y. Zhang, and T. Jiang, "A survey of underwater magnetic induction communications: Fundamental issues, recent advances, and challenges," *IEEE Communications Surveys & Tutorials*, vol. 21, no. 3, pp. 2466–2487, 2019.

[24] Z. Sun and I. F. Akyildiz, "Magnetic induction communications for wireless underground sensor networks," *IEEE transactions on antennas and propagation*, vol. 58, no. 7, pp. 2426–2435, 2010.

[25] T. E. Abrudan, Z. Xiao, A. Markham, and N. Trigoni, "Distortion rejecting magneto-inductive three-dimensional localization (MagLoc)," *IEEE Journal on Selected Areas in Communications*, vol. 33, no. 11, pp. 2404–2417, 2015.

[26] G. Dumphart, E. Slottke, and A. Wittneben, "Robust Near-Field 3D Localization of an Unaligned Single-Coil Agent Using Unobtrusive Anchors," in *IEEE PIMRC*, 2017.

[27] C. A. Balanis, *Antenna theory: analysis and design.* John wiley & sons, 2015.

[28] G. Dumphart, "Magneto-Inductive Communication and Localization: Fundamental Limits with Arbitrary Node Arrangements," Ph.D. dissertation, ETH Zurich, 2020.

[29] N. Golestani and M. Moghaddam, "Human activity recognition using magnetic induction-based motion signals and deep recurrent neural networks," *Nature communications*, vol. 11, no. 1, pp. 1–11, 2020.

[30] E. Shamonina, V. Kalinin, K. Ringhofer, and L. Solymar, "Magneto-inductive waveguide," *Electronics letters*, vol. 38, no. 8, pp. 371–373, 2002.

[31] E. Slottke, "Inductively coupled microsensor networks: Relay enabled cooperative communication and localization," Ph.D. dissertation, ETH Zurich, 2016.

[32] G. Dumphart, E. Slottke, and A. Wittneben, "Magneto-inductive Passive Relaying in Arbitrarily Arranged Networks," in *IEEE International Conference on Communications (ICC)*, May 2017. [Online]. Available: http://www.nari.ee.ethz.ch/wireless/pubs/p/PassiveRelaying

[33] C. K. Lee, W. X. Zhong, and S. Hui, "Effects of magnetic coupling of nonadjacent resonators on wireless power domino-resonator systems," *IEEE Transactions on Power Electronics*, vol. 27, no. 4, pp. 1905–1916, 2011.

[34] G. Dumphart, J. Sager, and A. Wittneben, "Load Modulation for Backscatter Communication: Channel Capacity and Capacity-Approaching Finite Constellations," *arXiv preprint arXiv:2207.08100*, 2022.

[35] O. Kypris, T. E. Abrudan, and A. Markham, "Magnetic induction-based positioning in distorted environments," *IEEE Transactions on Geoscience and Remote Sensing*, vol. 54, no. 8, pp. 4605–4612, 2016.

[36] P. Kela, M. Costa, J. Turkka, M. Koivisto, J. Werner, A. Hakkarainen, M. Valkama, R. Jantti, and K. Leppanen, "Location based beamforming in 5G ultra-dense networks," in *2016 IEEE 84th Vehicular Technology Conference (VTC-Fall)*. IEEE, 2016, pp. 1–7.

[37] P. Chen, Y. Yang, Y. Wang, and Y. Ma, "Adaptive beamforming with sensor position errors using covariance matrix construction based on subspace bases transition," *IEEE Signal Processing Letters*, vol. 26, no. 1, pp. 19–23, 2018.

[38] R. Heyn, M. Kuhn, H. Schulten, G. Dumphart, J. Zwyssig, F. Trosch, and A. Wittneben, "User Tracking for Access Control with Bluetooth Low Energy," in *VTC2019-Spring - IEEE 89th Vehicular Technology Conference*. IEEE, Apr. 2019, pp. 1–7.

[39] T. D. Than, G. Alici, H. Zhou, and W. Li, "A review of localization systems for robotic endoscopic capsules," *IEEE transactions on biomedical engineering*, vol. 59, no. 9, pp. 2387–2399, 2012.

[40] V. Pasku, A. De Angelis, G. De Angelis, D. D. Arumugam, M. Dionigi, P. Carbone, A. Moschitta, and D. S. Ricketts, "Magnetic field-based positioning systems," *IEEE Communications Surveys & Tutorials*, vol. 19, no. 3, pp. 2003–2017, 2017.

[41] G. Ouyang and K. Abed-Meraim, "A Survey of Magnetic-Field-Based Indoor Localization," *Electronics*, vol. 11, no. 6, p. 864, 2022.

[42] B. Li, T. Gallagher, A. G. Dempster, and C. Rizos, "How feasible is the use of magnetic field alone for indoor positioning?" in *2012 International Conference on Indoor Positioning and Indoor Navigation (IPIN)*. IEEE, 2012, pp. 1–9.

[43] Y. Shu, C. Bo, G. Shen, C. Zhao, L. Li, and F. Zhao, "Magicol: Indoor localization using pervasive magnetic field and opportunistic WiFi sensing," *IEEE Journal on Selected Areas in Communications*, vol. 33, no. 7, pp. 1443–1457, 2015.

[44] S. Song, C. Hu, and M. Q.-H. Meng, "Multiple objects positioning and identification method based on magnetic localization system," *IEEE Transactions on Magnetics*, vol. 52, no. 10, pp. 1–4, 2016.

[45] V. Schlageter, P.-A. Besse, R. Popovic, and P. Kucera, "Tracking system with five degrees of freedom using a 2D-array of Hall sensors and a permanent magnet," *Sensors and Actuators A: Physical*, vol. 92, no. 1-3, pp. 37–42, 2001.

[46] G. Shao and Y.-X. Guo, "Wearable magnetic localization system with noise cancellation for wireless capsule endoscopy," in *2019 IEEE MTT-S International Microwave Biomedical Conference (IMBioC)*, vol. 1. IEEE, 2019, pp. 1–3.

[47] C. Hu, Y. Ren, X. You, W. Yang, S. Song, S. Xiang, X. He, Z. Zhang, and M. Q.-H. Meng, "Locating intra-body capsule object by three-magnet sensing system," *IEEE Sensors Journal*, vol. 16, no. 13, pp. 5167–5176, 2016.

[48] J. Jung, S.-M. Lee, and H. Myung, "Indoor mobile robot localization and mapping based on ambient magnetic fields and aiding radio sources," *IEEE Transactions on Instrumentation and Measurement*, vol. 64, no. 7, pp. 1922–1934, 2014.

[49] A. Sheinker, B. Ginzburg, N. Salomonski, L. Frumkis, and B.-Z. Kaplan, "Localization in 3-D using beacons of low frequency magnetic field," *IEEE transactions on instrumentation and measurement*, vol. 62, no. 12, pp. 3194–3201, 2013.

[50] A. Sheinker, B. Ginzburg, N. Salomonski, and A. Engel, "Localization of a mobile platform equipped with a rotating magnetic dipole source," *IEEE Transactions on Instrumentation and Measurement*, vol. 68, no. 1, pp. 116–128, 2018.

[51] T. E. Abrudan, Z. Xiao, A. Markham, and N. Trigoni, "Underground incrementally deployed magneto-inductive 3-D positioning network," *IEEE Transactions on Geoscience and Remote Sensing*, vol. 54, no. 8, pp. 4376–4391, 2016.

[52] S. Kisseleff, X. Chen, I. F. Akyildiz, and W. Gerstacker, "Localization of a silent target node in magnetic induction based wireless underground sensor networks," in *2017 IEEE International Conference on Communications (ICC)*. IEEE, 2017, pp. 1–7.

[53] S.-C. Lin, A. A. Alshehri, P. Wang, and I. F. Akyildiz, "Magnetic induction-based localization in randomly deployed wireless underground sensor networks," *IEEE Internet of Things Journal*, vol. 4, no. 5, pp. 1454–1465, 2017.

[54] E. Slottke and A. Wittneben, "Circuit based near-field localization: calibration algorithms and experimental results," in *2015 IEEE 82nd Vehicular Technology Conference (VTC2015-Fall)*. IEEE, 2015, pp. 1–5.

[55] C. C. Finlay, S. Maus, C. Beggan, T. Bondar, A. Chambodut, T. Chernova, A. Chulliat, V. Golovkov, B. Hamilton, M. Hamoudi *et al.*, "International geomagnetic reference field: the eleventh generation," *Geophysical Journal International*, vol. 183, no. 3, pp. 1216–1230, 2010.

[56] F. Bianchi, A. Masaracchia, E. Shojaei Barjuei, A. Menciassi, A. Arezzo, A. Koulaouzidis, D. Stoyanov, P. Dario, and G. Ciuti, "Localization strategies for robotic endoscopic capsules: a review," *Expert review of medical devices*, vol. 16, no. 5, pp. 381–403, 2019.

[57] A. Markham, N. Trigoni, D. W. Macdonald, and S. A. Ellwood, "Underground localization in 3-D using magneto-inductive tracking," *IEEE Sensors Journal*, vol. 12, no. 6, pp. 1809–1816, 2011.

[58] Q. Huang, X. Zhang, and J. Ma, "Underground magnetic localization method and optimization based on simulated annealing algorithm," in *2015 IEEE 12th Intl Conf on Ubiquitous Intelligence and Computing and 2015 IEEE 12th Intl Conf on Autonomic and Trusted Computing and 2015 IEEE 15th Intl Conf on*

Scalable Computing and Communications and Its Associated Workshops (UIC-ATC-ScalCom). IEEE, 2015, pp. 168–173.

[59] X. Tan and Z. Sun, "Environment-aware indoor localization using magnetic induction," in *2015 IEEE Global Communications Conference (GLOBECOM).* IEEE, 2015, pp. 1–6.

[60] N. Zhang, H. Liu, S. Chen, M. Li, and Y. Jiang, "Indoor localization scheme for fire rescue based on super low frequency quasi-static field," in *2018 2nd IEEE Advanced Information Management, Communicates, Electronic and Automation Control Conference (IMCEC).* IEEE, 2018, pp. 1095–1099.

[61] G. Dumphart, H. Schulten, B. Bhatia, C. Sulser, and A. Wittneben, "Practical Accuracy Limits of Radiation-Aware Magneto-Inductive 3D Localization," in *ICC 2019 - IEEE International Conference on Communications Workshops.* IEEE, May 2019, pp. 1–6.

[62] S. Hashi, S. Yabukami, H. Kanetaka, K. Ishiyama, and K. Arai, "Numerical study on the improvement of detection accuracy for a wireless motion capture system," *IEEE Transactions on Magnetics*, vol. 45, no. 6, pp. 2736–2739, 2009.

[63] ——, "Wireless magnetic position-sensing system using optimized pickup coils for higher accuracy," *IEEE Transactions on Magnetics*, vol. 47, no. 10, pp. 3542–3545, 2011.

[64] E. Slottke and A. Wittneben, "Accurate localization of passive sensors using multiple impedance measurements," in *2014 IEEE 79th Vehicular Technology Conference (VTC Spring).* IEEE, 2014, pp. 1–5.

[65] ——, "Single-anchor localization in inductively coupled sensor networks using passive relays and load switching," in *2015 49th Asilomar Conference on Signals, Systems and Computers.* IEEE, 2015, pp. 214–218.

[66] P. Ltd. LIBERTY. [Online]. Available: https://polhemus.com/motion-tracking/all-trackers/liberty

[67] M. J. Wheare, M. J. Nelson, R. Lumsden, A. Buttfield, and R. G. Crowther, "Reliability and Validity of the Polhemus Liberty System for Upper Body Segment and Joint Angular Kinematics of Elite Golfers," *Sensors*, vol. 21, no. 13, p. 4330, 2021.

[68] X. Huang, F. Wang, J. Zhang, Z. Hu, and J. Jin, "A posture recognition method based on indoor positioning technology," *Sensors*, vol. 19, no. 6, p. 1464, 2019.

[69] Z. W. Mekonnen, "Time of Arrival Based Infrastructureless Human Posture Capturing System," Ph.D. dissertation, ETH Zurich, 2016.

[70] G. Rogez, J. Rihan, S. Ramalingam, C. Orrite, and P. H. Torr, "Randomized trees for human pose detection," in *2008 IEEE Conference on Computer Vision and Pattern Recognition.* IEEE, 2008, pp. 1–8.

[71] K. M. Sagayam and D. J. Hemanth, "Hand posture and gesture recognition techniques for virtual reality applications: a survey," *Virtual Reality*, vol. 21, no. 2, pp. 91–107, 2017.

[72] M. Yu, A. Rhuma, S. M. Naqvi, L. Wang, and J. Chambers, "A posture recognition-based fall detection system for monitoring an elderly person in a smart home environment," *IEEE transactions on information technology in biomedicine*, vol. 16, no. 6, pp. 1274–1286, 2012.

[73] R. Heyn and A. Wittneben, "Detection of Fall-Related Body Postures from WBAN Signals," in *GLOBECOM 2020-2020 IEEE Global Communications Conference.* IEEE, 2020, pp. 1–6.

[74] L. Simpson, M. M. Maharaj, and R. J. Mobbs, "The role of wearables in spinal posture analysis: a systematic review," *BMC musculoskeletal disorders*, vol. 20, no. 1, pp. 1–14, 2019.

[75] H. Zhou, Y. Gao, W. Liu, Y. Jiang, and W. Dong, "Posture Tracking Meets Fitness Coaching: A Two-Phase Optimization Approach with Wearable Devices," in *2020 IEEE 17th International Conference on Mobile Ad Hoc and Sensor Systems (MASS).* IEEE, 2020, pp. 524–532.

[76] A. M. Briggs *et al.*, "Musculoskeletal health conditions represent a global threat to healthy aging: a report for the 2015 world health organization world report on ageing and health," *The Gerontologist*, vol. 56, no. 2, pp. 243–255, 2016.

[77] M. Dimitrijevic, V. Lepetit, and P. Fua, "Human body pose detection using bayesian spatio-temporal templates," *Computer vision and image understanding*, vol. 104, no. 2-3, pp. 127–139, 2006.

[78] R. Heyn and A. Wittneben, "Comprehensive Measurement-Based Evaluation of Posture Detection from Ultra Low Power UWB Signals," in *2021 IEEE 32nd Annual International Symposium on Personal, Indoor and Mobile Radio Communications (PIMRC)*. IEEE, 2021, pp. 1518–1524.

[79] C.-C. Yang and Y.-L. Hsu, "A review of accelerometry-based wearable motion detectors for physical activity monitoring," *Sensors*, vol. 10, no. 8, pp. 7772–7788, 2010.

[80] A. Crane, S. Doppalapudi, J. O'Leary, P. Ozarek, and C. Wagner, "Wearable posture detection system," in *2014 40th Annual Northeast Bioengineering Conference (NEBEC)*. IEEE, 2014, pp. 1–2.

[81] E. Farella, A. Pieracci, L. Benini, and A. Acquaviva, "A wireless body area sensor network for posture detection," in *11th IEEE Symposium on Computers and Communications (ISCC'06)*. IEEE, 2006, pp. 454–459.

[82] S. Q. Liu, J. C. Zhang, and R. Zhu, "A wearable human motion tracking device using micro flow sensor incorporating a micro accelerometer," *IEEE Transactions on Biomedical Engineering*, vol. 67, no. 4, pp. 940–948, 2019.

[83] R. Gupta, S. H. Gupta, A. Agarwal, P. Choudhary, N. Bansal, and S. Sen, "A Wearable Multisensor Posture Detection System," in *2020 4th International Conference on Intelligent Computing and Control Systems (ICICCS)*. IEEE, 2020, pp. 818–822.

[84] H. Harms, O. Amft, G. Troester, M. Appert, R. Mueller, and A. Meyer-Heim, "Wearable therapist: sensing garments for supporting children improve posture," in *Proceedings of the 11th international conference on Ubiquitous computing*, 2009, pp. 85–88.

[85] Q. Wang, W. Chen, A. A. Timmermans, C. Karachristos, J.-B. Martens, and P. Markopoulos, "Smart Rehabilitation Garment for posture monitoring," in *2015 37th annual International Conference of the IEEE engineering in medicine and biology Society (EmbC)*. IEEE, 2015, pp. 5736–5739.

[86] C. Mattmann, "Body posture detection using strain sensitive clothing," Ph.D. dissertation, ETH Zurich, 2008.

[87] E. Sardini, M. Serpelloni, and M. Ometto, "Smart vest for posture monitoring in rehabilitation exercises," in *2012 IEEE Sensors Applications Symposium Proceedings.* IEEE, 2012, pp. 1–5.

[88] S. Wielgos, E. Dolezalek, and C.-H. Min, "Garment Integrated Spinal Posture Detection Using Wearable Magnetic Sensors," in *2020 42nd Annual International Conference of the IEEE Engineering in Medicine & Biology Society (EMBC).* IEEE, 2020, pp. 4030–4033.

[89] H. Schulten and A. Wittneben, "Robust Multi-Frequency Posture Detection based on Purely Passive Magneto-Inductive Tags," in *ICC 2022 - IEEE International Conference on Communications.* IEEE, May 2022, pp. 1–6.

[90] ——, "Experimental Study of Posture Detection Using Purely Passive Magneto-Inductive Tags," in *WCNC 2022 - IEEE Wireless Communications and Networking Conference.* IEEE, Apr. 2022, pp. 1–6.

[91] H. Schulten, F. Wernli, and A. Wittneben, "Learning-Based Posture Detection Using Purely Passive Magneto-Inductive Tags," in *Globecom 2021 - IEEE Global Communications Conference.* IEEE, Dec. 2021, pp. 1–6.

[92] H. Schulten, G. Dumphart, A. Koskinas, and A. Wittneben, "Cooperative Magneto-Inductive Localization," in *PIMRC 2021 - IEEE 32nd Annual International Symposium on Personal, Indoor and Mobile Radio Communications.* IEEE, Sep. 2021, pp. 1–7.

[93] H. Schulten and A. Wittneben, "Magneto-Inductive Localization: Fundamentals of Passive Relaying and Load Switching," in *ICC 2020 - IEEE International Conference on Communications.* IEEE, Jun. 2020, pp. 1–6.

[94] H. Schulten, M. Kuhn, R. Heyn, G. Dumphart, F. Trosch, and A. Wittneben, "On the crucial impact of antennas and diversity on BLE RSSI-based indoor localization," in *VTC2019-Spring - IEEE 89th Vehicular Technology Conference.* IEEE, Apr. 2019, pp. 1–6.

[95] H. Schulten and A. Wittneben, "Method and Apparatus for Determining a Spatial Configuration of a Wireless Inductive Network and for Pose Detection," Europe Patent Request EP21 211 962.2, Dec., 2021.

[96] F. Trosch, A. Wittneben, H. Schulten, J. Zwyssig, and M. Kuhn, "Zugangskontrollsystem und Verfahren zum Betreiben eines Zugangskontrollsystems," World Patent WO2 020 216 877A1, Oct., 2020.

[97] M. T. Ivrlac and J. A. Nossek, "Toward a circuit theory of communication," *IEEE Transactions on Circuits and Systems I: Regular Papers*, vol. 57, no. 7, pp. 1663–1683, 2010.

[98] J. C. Maxwell, "On physical lines of force," *The London, Edinburgh, and Dublin Philosophical Magazine and Journal of Science*, vol. 21, no. 139, pp. 161–175, 1861.

[99] ——, "A dynamical theory of the electromagnetic field (1865)," *The Scientific Papers of James Clerk Maxwell*, vol. 2, 1890.

[100] T. H. Hubing, "Survey of numerical electromagnetic modeling techniques," *Department of Electrical Engineering, University of Missouri-Rolla, USA*, 1991.

[101] M. B. Oezakin and S. Aksoy, "Application of magneto-quasi-static approximation in the finite difference time domain method," *IEEE Transactions on Magnetics*, vol. 52, no. 8, pp. 1–9, 2016.

[102] R. P. Feynman, R. B. Leighton, and M. Sands, *The Feynman Lectures on Physics (The New Millenium Edition) Volume II: Mainly Electromagnetism and Matter.* Basic Books, 2011, vol. 2, available at www.feynmanlectures.caltech.edu.

[103] J. E. Storer, "Impedance of thin-wire loop antennas," *Transactions of the American Institute of Electrical Engineers, Part I: Communication and Electronics*, vol. 75, no. 5, pp. 606–619, 1956.

[104] F. E. Neumann, "Allgemeine Gesetze der inducirten elektrischen Stroeme," *Annalen der Physik*, vol. 143, no. 1, pp. 31–44, 1846.

[105] D. Eberly, "Euler angle formulas," *Geometric Tools, LLC, Technical Report*, pp. 1–18, 2008.

[106] S. Kisseleff, I. F. Akyildiz, and W. Gerstacker, "Interference polarization in magnetic induction based wireless underground sensor networks," in *2013 IEEE 24th International Symposium on Personal, Indoor and Mobile Radio Communications (PIMRC Workshops).* IEEE, 2013, pp. 71–75.

[107] S. Kisseleff, I. F. Akyildiz, and W. H. Gerstacker, "Throughput of the magnetic induction based wireless underground sensor networks: Key optimization techniques," *IEEE Transactions on Communications*, vol. 62, no. 12, pp. 4426–4439, 2014.

[108] K. Finkenzeller, *RFID handbook: fundamentals and applications in contactless smart cards, radio frequency identification and near-field communication.* John wiley and sons, 2010.

[109] R. Dengler, "Self inductance of a wire loop as a curve integral," *arXiv preprint arXiv:1204.1486*, 2012.

[110] D. W. Knight, "The self-resonance and self-capacitance of solenoid coils," 2016, DOI: 10.13140/RG.2.1.1472.0887. [Online]. Available: http://g3ynh.info/zdocs/magnetics/appendix/self_res/self-res.pdf

[111] R. Medhurst, "HF resistance and self-capacitance of single-layer solenoids," *Wireless Engineer*, vol. 24, 1947.

[112] S. Phang, M. T. Ivrlac, G. Gradoni, S. C. Creagh, G. Tanner, and J. A. Nossek, "Near-field MIMO communication links," *IEEE Transactions on Circuits and Systems I: Regular Papers*, vol. 65, no. 9, pp. 3027–3036, 2018.

[113] Y. Hassan, "Compact Multi-Antenna Systems: Bridging Circuits to Communications Theory," Ph.D. dissertation, PhD Thesis, ETH Zurich, May 2018.

[114] D. Nie, B. M. Hochwald, and E. Stauffer, "Systematic design of large-scale multiport decoupling networks," *IEEE Transactions on Circuits and Systems I: Regular Papers*, vol. 61, no. 7, pp. 2172–2181, 2014.

[115] S. M. Kay, *Fundamentals of statistical signal processing: estimation theory.* Prentice-Hall, Inc., 1993.

[116] A. S. Poznyak, *Advanced mathematical tools for automatic control engineers: Stochastic techniques.* Elsevier, 2009.

[117] E. A. Carlen, "Superadditivity of Fisher's information and logarithmic Sobolev inequalities," *Journal of Functional Analysis*, vol. 101, no. 1, pp. 194–211, 1991.

[118] Y. Shen, H. Wymeersch, and M. Z. Win, "Fundamental limits of wideband localization—Part II: Cooperative networks," *IEEE Transactions on Information Theory*, vol. 56, no. 10, pp. 4981–5000, 2010.

[119] D. B. Jourdan and N. Roy, "Optimal sensor placement for agent localization," *ACM Transactions on Sensor Networks (TOSN)*, vol. 4, no. 3, pp. 1–40, 2008.

[120] N. Levanon, "Lowest GDOP in 2-D scenarios," *IEE Proceedings-radar, sonar and navigation*, vol. 147, no. 3, pp. 149–155, 2000.

[121] J. T. Isaacs, D. J. Klein, and J. P. Hespanha, "Optimal sensor placement for time difference of arrival localization," in *Proceedings of the 48h IEEE Conference on Decision and Control (CDC) held jointly with 2009 28th Chinese Control Conference.* IEEE, 2009, pp. 7878–7884.

[122] A. Kurs, A. Karalis, R. Moffatt, J. D. Joannopoulos, P. Fisher, and M. Soljacic, "Wireless power transfer via strongly coupled magnetic resonances," *science*, vol. 317, no. 5834, pp. 83–86, 2007.

[123] Z. Sun, I. F. Akyildiz, S. Kisseleff, and W. Gerstacker, "Increasing the capacity of magnetic induction communications in RF-challenged environments," *IEEE Transactions on Communications*, vol. 61, no. 9, pp. 3943–3952, 2013.

[124] S. Kisseleff, W. Gerstacker, R. Schober, Z. Sun, and I. F. Akyildiz, "Channel capacity of magnetic induction based wireless underground sensor networks under practical constraints," in *2013 IEEE Wireless Communications and Networking Conference (WCNC).* IEEE, 2013, pp. 2603–2608.

[125] Y. Hassan, B. Gahr, and A. Wittneben, "Rate maximization in dense interference networks using non-cooperative passively loaded relays," in *2015 49th Asilomar Conference on Signals, Systems and Computers.* IEEE, 2015, pp. 978–982.

[126] H. Wymeersch, J. Lien, and M. Z. Win, "Cooperative localization in wireless networks," *Proceedings of the IEEE*, vol. 97, no. 2, pp. 427–450, 2009.

[127] N. Patwari, J. N. Ash, S. Kyperountas, A. O. Hero, R. L. Moses, and N. S. Correal, "Locating the nodes: cooperative localization in wireless sensor networks," *IEEE Signal processing magazine*, vol. 22, no. 4, pp. 54–69, 2005.

[128] M. Z. Win, A. Conti, S. Mazuelas, Y. Shen, W. M. Gifford, D. Dardari, and M. Chiani, "Network localization and navigation via cooperation," *IEEE Communications Magazine*, vol. 49, no. 5, pp. 56–62, 2011.

[129] P. Zhang, J. Lu, Y. Wang, and Q. Wang, "Cooperative localization in 5G networks: A survey," *Ict Express*, vol. 3, no. 1, pp. 27–32, 2017.

[130] G. Dumphart and A. Wittneben, "Stochastic misalignment model for magneto-inductive SISO and MIMO links," in *2016 IEEE 27th Annual International Symposium on Personal, Indoor, and Mobile Radio Communications (PIMRC)*. IEEE, 2016, pp. 1–6.

[131] J. Zhou and J. Shi, "RFID localization algorithms and applications—a review," *Journal of intelligent manufacturing*, vol. 20, no. 6, pp. 695–707, 2009.

[132] P. Tarrío, A. M. Bernardos, and J. R. Casar, "Weighted least squares techniques for improved received signal strength based localization," *Sensors*, vol. 11, no. 9, pp. 8569–8592, 2011.

[133] L. Mirsky, "A trace inequality of John von Neumann," *Monatshefte fuer mathematik*, vol. 79, no. 4, pp. 303–306, 1975.

[134] P. H. Schoenemann, "A generalized solution of the orthogonal procrustes problem," *Psychometrika*, vol. 31, no. 1, pp. 1–10, 1966.

[135] W. Kabsch, "A solution for the best rotation to relate two sets of vectors," *Acta Crystallographica Section A: Crystal Physics, Diffraction, Theoretical and General Crystallography*, vol. 32, no. 5, pp. 922–923, 1976.

[136] K. Levenberg, "A method for the solution of certain non-linear problems in least squares," *Quarterly of applied mathematics*, vol. 2, no. 2, pp. 164–168, 1944.

[137] *1 MSPS, 12-Bit Impedance Converter, Network Analyzer - Data Sheet AD5933*, Analog Devices, Inc., 2017. [Online]. Available: https://www.analog.com/media/en/technical-documentation/data-sheets/AD5933.pdf

[138] *Impedance Measurement Handbook A Guide to Measurement Technology and Techniques*, Keysight Technologies, 6 2020.

[139] H. Fischer, *A history of the central limit theorem: from classical to modern probability theory*. Springer, 2011.

[140] A. Goldsmith, *Wireless communications.* Cambridge university press, 2005.

[141] C. M. Bishop, *Pattern recognition and machine learning.* Springer, 2006.

[142] R. B. Cattell, "The scree test for the number of factors," *Multivariate behavioral research*, vol. 1, no. 2, pp. 245–276, 1966.

[143] W. Gilchrist, *Statistical modelling with quantile functions.* Chapman and Hall/CRC, 2000.

[144] B. Sun, J. Feng, and K. Saenko, "Return of frustratingly easy domain adaptation," in *Proceedings of the AAAI Conference on Artificial Intelligence*, vol. 30, no. 1, 2016.

[145] J. Gu, Z. Wang, J. Kuen, L. Ma, A. Shahroudy, B. Shuai, T. Liu, X. Wang, G. Wang, J. Cai *et al.*, "Recent advances in convolutional neural networks," *Pattern recognition*, vol. 77, pp. 354–377, 2018.

[146] Y. Yu, X. Si, C. Hu, and J. Zhang, "A review of recurrent neural networks: LSTM cells and network architectures," *Neural computation*, vol. 31, no. 7, pp. 1235–1270, 2019.

[147] L. Van der Maaten and G. Hinton, "Visualizing data using t-SNE," *Journal of machine learning research*, vol. 9, no. 11, 2008.

[148] D. A. Reynolds, "Gaussian mixture models," *Encyclopedia of biometrics*, vol. 741, no. 659-663, 2009.

[149] D. Amodei, C. Olah, J. Steinhardt, P. Christiano, J. Schulman, and D. Mané, "Concrete problems in AI safety," *arXiv preprint arXiv:1606.06565*, 2016.

[150] K.-i. Kanatani, "Analysis of 3-D rotation fitting," *IEEE Transactions on pattern analysis and machine intelligence*, vol. 16, no. 5, pp. 543–549, 1994.

Bisher erschienene Bände der Reihe

Series in Wireless Communications

ISSN 1611-2970

1	Dirk Benyoucef	Codierte Multiträgerverfahren für Powerline- und Mobilfunkanwendungen ISBN 978-3-8325-0137-2 40.50 EUR
2	Marc Kuhn	Space-Time Codes und ihre Anwendungen in zukünftigen Kommunikationssystemen ISBN 978-3-8325-0159-4 40.50 EUR
3	Frank Althaus	A New Low-Cost Approach to Wireless Communication over Severe Multipath Fading Channels ISBN 978-3-8325-0158-7 40.50 EUR
4	Ingmar Hammerström	Cooperative Relaying and Adaptive Scheduling for Low Mobility Wireless Access Networks ISBN 978-3-8325-1466-2 40.50 EUR
5	Thomas Zasowski	A System Concept for Ultra Wideband (UWB) Body Area Networks ISBN 978-3-8325-1715-1 40.50 EUR
6	Boris Rankov	Spectral Efficient Cooperative Relaying Strategies for Wireless Networks ISBN 978-3-8325-1744-1 40.50 EUR
7	Stephan Sand	Joint Iterative Channel and Data Estimation in High Mobility MIMO-OFDM Systems ISBN 978-3-8325-2385-5 39.00 EUR
8	Florian Troesch	Design and Optimization of Distributed Multiuser Cooperative Wireless Networks ISBN 978-3-8325-2400-5 42.50 EUR
9	Celal Esli	Novel Low Duty Cycle Schemes: From Ultra Wide Band to Ultra Low Power ISBN 978-3-8325-2484-5 42.00 EUR

10	Stefan Berger	Coherent Cooperative Relaying in Low Mobility Wireless Multiuser Networks ISBN 978-3-8325-2536-1 42.00 EUR
11	Christoph Steiner	Location Fingerprinting for Ultra-Wideband Systems. The Key to Efficient and Robust Localization ISBN 978-3-8325-2567-5 36.00 EUR
12	Jian Zhao	Analysis and Design of Communication Techniques in Spectrally Efficient Wireless Relaying Systems ISBN 978-3-8325-2585-9 39.50 EUR
13	Azadeh Ettefagh	Cooperative WLAN Protocols for Multimedia Communication ISBN 978-3-8325-2774-7 37.50 EUR
14	Georgios Psaltopoulos	Affordable Nonlinear MIMO Systems ISBN 978-3-8325-2824-9 36.50 EUR
15	Jörg Wagner	Distributed Forwarding in Multiuser Multihop Wireless Networks ISBN 978-3-8325-3193-5 36.00 EUR
16	Etienne Auger	Wideband Multi-user Cooperative Networks: Theory and Measurements ISBN 978-3-8325-3209-3 38.50 EUR
17	Heinrich Lücken	Communication and Localization in UWB Sensor Networks. A Synergetic Approach ISBN 978-3-8325-3332-8 36.50 EUR
18	Raphael T. L. Rolny	Future Mobile Communication: From Cooperative Cells to the Post-Cellular Relay Carpet ISBN 978-3-8325-4229-0 42.00 EUR
19	Zemene Walle Mekonnen	Time of Arrival Based Infrastructureless Human Posture Capturing System ISBN 978-3-8325-4429-4 36.00 EUR
20	Eric Slottke	Inductively Coupled Microsensor Networks: Relay Enabled Cooperative Communication and Localization ISBN 978-3-8325-4438-6 38.00 EUR

21	Tim Rüegg	Low Complexity Physical Layer Cooperation Concepts for Mobile Ad Hoc Networks ISBN 978-3-8325-4801-8 36.00 EUR
22	Gregor Dumphart	Magneto-Inductive Communication and Localization: Fundamental Limits with Arbitrary Node Arrangements ISBN 978-3-8325-5483-5 49.50 EUR
23	Henry Ruben Lucas Schulten	Localization and Posture Recognition via Magneto-Inductive and Relay-Aided Sensor Networks ISBN 978-3-8325-5586-3 50.50 EUR